A Chance for Change

Sharecropper to Landowner

The Histo-Autobiography of J. R. Rosser, Sr.

Compiled by J. R. Rosser, Jr.

Edited by Audrey Rosser Milo

Copyright © 2009 by Audrey Rosser Milo.

Library of Congress Control Number: 2009904153
ISBN: Hardcover   978-1-4415-3233-6
             Softcover   978-1-4415-3232-9

All rights reserved. No part of this book may be reproduced or transmitted in any form or by any means, electronic or mechanical, including photocopying, recording, or by any information storage and retrieval system, without permission in writing from the copyright owner.

This book was printed in the United States of America.

To order additional copies of this book, contact:
Xlibris Corporation
1-888-795-4274
www.Xlibris.com
Orders@Xlibris.com
58775

Contents

PREFACE.. 11
ACKNOWLEDGMENTS....................................... 13
INTRODUCTION.. 15

PART ONE: MEMOIRS

LONE OAK, GEORGIA, C. 1813-1906 18
        Ancestors
    Chapter 1 Pap Dean................................. 19
    Chapter 2 Grandparents............................. 22
    Chapter 3 Parents.................................. 25
        Endnotes................................. 29

MERIWETHER COUNTY, 1906-1923 31
        Chronology
    Chapter 4 Early Years............................... 32
    Chapter 5 School and Church....................... 37
    Chapter 6 Growing Up.............................. 40
    Chapter 7 Same Name............................... 44
    Chapter 8 Farm Life................................ 48
        Endnotes................................. 53

FORSYTH AND ATLANTA, 1923-1931 55
        Chronology
    Chapter 9   Boarding School.......................... 56
    Chapter 10 High School............................. 63
    Chapter 11 First Teaching Job........................ 68
    Chapter 12 AURS.................................... 70
        Endnotes................................. 73

HARALSON COUNTY, 1931-1946 75
        Chronology
    Chapter 13 Bremen and Waco........................ 76
    Chapter 14 Marriage................................ 78
    Chapter 15 Tallapoosa.............................. 83
    Chapter 16 Straight and Narrow..................... 91
    Chapter 17 More Education.......................... 94
        Endnotes................................. 95

CARROLL COUNTY, 1946-1950 97
            Chronology
  Chapter 18 Alternatives 98
  Chapter 19 A Full Load 101
            Endnotes 104
CHATTOOGA COUNTY, 1950-1958 108
            Chronology
  Chapter 20 Moving Ahead 109
  Chapter 21 Signs of the Times 114
  Chapter 22 GIA 116
  Chapter 23 Unwritten Rules 118
            Endnotes 119
HABERSHAM COUNTY, 1958-1970 121
            Chronology
  Chapter 24 Regional School 122
  Chapter 25 Racial Residue 125
  Chapter 26 Retirement 130
            Endnotes 132
GOLDEN YEARS, 1970-2004 133
            Chronology
  Chapter 27 Chamber of Commerce 135
  Chapter 28 A Good Citizen 137
            Endnotes 140

PART TWO: REFLECTIONS

INTERVIEWS
  Chapter 29 Mama's Influence 145
  Chapter 30 Twentieth-Century Milestones 149
  Chapter 31 The Human Condition 157
  Chapter 32 Lessons Learned 160
LAST WORD 164

PART THREE: QUOTATIONS

SCRIPTURE 169
POETRY 175
ADVICE 191
ONE LINERS 193
LYRICS 195

DEDICATION

In loving memory of Momma

Purpose

I observed my ninety-fourth birthday on Saturday, December 30, 2000. I want to record how my presence on earth and my actions make a difference. I do not know if my memory can satisfy the kind of detail one may want to know. I do not have very much written about it, and my memory is kind of dull. The Lord has spared me to be sound of mind, which allows this joint undertaking with the assistance of Robert to explore the meaning of life.

This writing is intended to provide an account of significant aspects of my life that suggest the fulfillment of a purposeful life. How did my presence and actions make a difference? What are the experiences that document my presence and actions? What lessons were learned? What issues remain?

The effort is being undertaken as my autobiography. It is intended for family, my own descendants and relatives, and descendants of relatives; former students and teachers and their descendants; friends; elected officials, local, state, and federal; and people in communities where I lived.

J. R. Rosser, Sr.
March 22, 2001

Preface

This work is the memoirs and reflections of John Robert Rosser, Sr., one Black man's life in the twentieth century, USA. The manuscript comes from notes and recordings of conversations with his eldest son, John R. Rosser, Jr., and his youngest daughter, Audrey Rosser Milo. This idea began December 29, 2000, following the death on December 15 of Pauline Lynch Rosser. The effort, launched January 2001, goes back throughout their lives based on stories often heard, people sometimes known, and places visited that he talked about on frequent occasions.

The approach was to examine his experiences from a geographical and chronological perspective using three themes: (1) Experiences that document his presence and actions. (2) Lessons of life discerned from the facts. (3) Issues and challenges remaining for others. We gathered information about theme 1 for each phase of his life; then for each phase, he considered themes 2 and 3 with general conclusions.

At age ninety-six he is affected by three centuries of life as a Black male in the United States of America. Being born in the first decade of the twentieth century, his early childhood was shaped by the customs and traditions in the nineteenth century that were set forth for Black people born in slavery and living forty years in the shadow of the Emancipation Proclamation.

His education, marriage, child rearing, writings, and retirement covered the twentieth century with its gamut of changes through the industrial, political, technological, and cultural explosion that he witnessed. He moved boldly into the twenty-first century in the twilight of his years still imbued with devotion to God and service to mankind through education and community service, still, open to the differences he might make in the life of someone else.

On December 28, 1996, at the Holiday Inn Central in Atlanta, Georgia, Dad's ninetieth birthday celebration was held. Among the tributes distributed that evening was a sixty-six-page monograph titled "Nearly Ninety: The Life and Times of John R. Rosser" by Myrna Suejette Rosser Riley, the elder of his two daughters. Where appropriate, endnotes will reference this work.

J. Robert Rosser, Jr., EdD
July 26, 2003

Acknowledgments

Pauline Lynch Rosser, wife of J. R. Rosser, Sr.

Because of who she was, what she did, when she spoke, where she kept silent, how she lived; and without whom, this book would not have been possible.

Myrna Suejette Rosser Riley

The first daughter, who spent countless hours through the years asking for and listening to Dad's life stories. Her suggestions to include stories that she often heard and compiled significantly augmented explanatory facts that Dad omitted retelling from his last revelations. Additionally, our sister contributed various related details that emerged from conversations with both of our parents. We gratefully acknowledge that her input confirmed, clarified, and enhanced previous information.

Pearl Lockhart Rosser, MD, wife of S. B. Rosser, MD

Whose suggestions for organization of content were invaluable and shaped the readability of the stories. Her sharp editing skills and insightful choices were key to the final selections. She remained involved and supportive until her untimely death, November 2008.

Florence Towles Rosser, wife of J. R. Rosser, Jr.

Her contributions for historical accuracy were an important constituent to early drafts. Editing assistance provided prior to her illness and death in January 2008 was helpful and deeply appreciated.

The Publishing Staff

Profound gratitude to Charles Andre, Lynnel Landerito, Charliz Elle, Michelle Postrano, and all other editors for their patience, kindness, and expertise rendered at each level of the process.

Introduction

Every autobiography is historical to the extent that it is an experiential artifact representing a specific time. Because of the impact of three centuries of mores on his life, these recollections by my father are recorded as a histo-autobiography. The chronicle of his life with national and international events and his responses to the circumstances, like any history, is potentially a predictor of the future and of possibility; however, countless Americans from the same era may not have shared the same outcomes as those produced by his integrity.

The mosaic of his life reflects behaviors that were realities for the times in which he lived. They express a predilection and philosophy that are both written and unwritten. Readers should be forewarned of the frankness with which he speaks. He has told the truth as he saw it, felt it, and knew it. He earned that right.

To those who do not share his opinions, he would have been the first to defend your right to your own perspectives. He expressed the preference that this be published posthumously. I believe it would have been unfair to ask him to defend it, but he would have welcomed the debate. As you read, I invite you to look for his affirmations and declarations that are interwoven between the lines. These attest to lessons he learned and serve as admonitions to others.

Occasionally he uses words and terms that may seem archaic and arcane; however, they were the norm during the times he learned them. Ergo, they affirm the maintenance of historic authenticity. An example of what may be regarded as Ebonics today occurs with the words *sometime* and *folk*. Neither of them has an *s* regardless to how it is used.

To every extent possible, in the time frame for editing, the correct spelling of names and the accuracy of dates have been researched in organizational records, library archives, school records, courthouse records, cemetery headstones, published maps, Web sites, and personal conversations.

He uses actual names because this is a work of nonfiction. No attempt has been made to protect the innocent, excuse the ignorant, or prosecute the guilty. In instances where his memory of someone may not be favorable, please be assured that no malice of forethought, hurt, or embarrassment was intended to the person, to their memory, or to any relatives or acquaintances.

This content was gathered the last three and a half years of Dad's life. It began with his stories on tape and evolved the last eighteen months to note taking from oral explanations and significant details as he spontaneously remembered them.

The two data-collection processes, conversation and interview, preserve the same sentence structure that he used, to every extent possible for the sake of clarity. Occasionally sentences are added from papers, letters, articles, and some of his earlier autobiographical accounts that were discovered after his death.

The word *Black* is capitalized throughout this discourse when the references are to culture and race. It is not merely a descriptor of skin color; it is a state of consciousness. It is an accepted designation for people of African descent who live in the United States of America, and is used as frequently as the name *African American*.

In *Black History and Black Identity: A Call for a New Historiography* (2001), W. D. Wright discusses the choice to capitalize the first letter of *Black* as a designation of ethnicity and I agree with the rationale. It is also applied to the use of *Colored*. These are exceptions to *The Chicago Manual of Style* used as the publishing standard; however, it is my interpretation of the value and respect Dad felt about who he was. This respect is also extended to other cultures, which is why the word *White* is capitalized.

Conversations with Minnie Ellis Freeman and a copy of her unpublished *Family Tree & Diagram of the St. Paul CME Church of Lone Oak, Georgia in Meriwether County* (1985) were valuable resources. For two years she and her son, Harold Freeman, compiled names of ascendants and descendants. They printed copies for limited distribution to these families. All other writings, published and unpublished, from which information is quoted are referenced after each section in endnotes.

Dad was a much sharper source than he gave himself credit for being. His ability to recall names, places, and quotes was remarkable. The two most influential women in his life were unequivocally his mother and his wife. The references to his mother are always *momma*, while references to his wife are always *mama*. This distinction is intentional.

After numerous readings, I remain in awe of his life. In summary, it was the determination for a *chance* to have a formal education that left a legacy of *change* through a strong charge for moral excellence. He often recounted the fact that when he was born his chances for success were "slim to none." He lived to overcome those odds.

After contracting pneumonia in February and a heart attack in April, he slowed considerably, but his indomitable spirit remained steadfast until his peaceful move to eternal rest in May. The four and a half years of editing have been a slow and painstaking process flooded by remembrance. There may be some lack of continuity from what he envisioned eight years ago, but in the words of a man he greatly admired, "It is finished."

Audrey Rosser Milo, EdD
January 30, 2009

Part One

Memoirs

LONE OAK, GEORGIA

Name	Birth	Death
Peter Dean Rosser	Circa 1813	Circa 1926
John Irvin Rosser	January 14, 1852	June 21, 1930
John Taylor Barrow	November 15, 1853	November 19, 1939
Elmire Smith Rosser	March 12, 1857	May 29, 1929
William Ellis	March 20, 1859	October 15, 1900
USA Civil War	*April 12, 1862*	*May 26, 1865*
Jennie Hutchin Ellis	October 15, 1865	November 7, 1924
Robert Miles Rosser	August 29, 1881	April 24, 1964
Caroline Ellis Rosser	February 28, 1882	July 20, 1931

Chapter 1

PAP DEAN

My great-grandfather had two surnames. He was called Peter Rosser sometime and Peter Dean sometime. He was in the Civil War. While he was gone, his wife and children were sold to the Rosser family along with the property. When he came back, we were Rossers, instead of Deans.

I don't know how he got into the army; he didn't tell me much about that. He was a slave, and I suspect his folk were sold to somebody else, but he was kept there by the Dean[1] man, a slave master. I'd guess he was manservant to the young Dean fellow, so when his master went off to war, he went with him.

He was the oldest man around the community, and everybody called him Pap. He had a beard like Santa Claus and favored Alfrez Shuffer[2] in hair, beard, and build. He was sort of short, average, about like I am. I'm not the most "dubbed off" fellow you've seen. I'm shorter now, but I was five feet eight inches tall.

Pap was half-Cherokee Indian and half-White. His mother was a Cherokee Indian woman. His father was an unidentified White man who raped his mother and left her for dead by the side of the road. A slave family that went by the name of Dean rescued her. They took her in and took care of her for the rest of her life.

From that ordeal, she turned out to be pregnant. That was a common problem among Black womenfolk at that time. Those White men did whatever they wanted with any of the women there. That baby she was carrying turned out to be Pap.

I never heard folk say what his mother's name was. But I remember seeing her a time or two when I was a little boy. She had two long braids of hair hanging down. She never said a word. Every time I saw her, she was sitting in a rocking chair and had something on her lap that she was working on. Some kind of handicraft. They didn't let me talk to her, so I don't know anything else about her.

Pap's Indian ancestors came down through Lone Oak to a spring where there used to be two oak trees. Once a year the Cherokees and the Seminoles would meet there because they were friends. After lightning struck and killed one of the oaks, the place was called "the lone oak," and that's how the town got its name.

Pap was considered a Negro and a slave, but he was a cook in the Confederate army. A whole lot of Blacks went into the army. At that time the soldiers got a little money. They had to have snuff. He told me that when the war was over, he had seventy-five dollars to "die" on him.

It was that Confederate money from the Confederate army. After the war, it wasn't worth anything, about like Continental money after the Revolutionary War wasn't worth much. That's where the saying "It ain't worth a continental" came from.

When Pap was 100 years old, he told me this story. It happened about 1865. Pap was in the Battle of Chickamauga, up above Atlanta, close to Chattanooga, Tennessee. He was a sergeant and was appointed bodyguard of a captain. The big battle was about over. The Union was destroying his army. The captain sent what men he had left, fleeing southward. For some cause I don't know, the captain was too sick to go, so Pap had to stay there and watch over him. Their camp was near Chickamauga.

Sherman came through Georgia. He burned Atlanta and went on to Savannah. After Sherman and his army moved on in his route to the sea, a Union cleanup crew was coming behind them and came into their particular campsite. Pap hid the captain under a pile of dirty uniforms. The soldiers quizzed him and knocked him around, but he convinced them that nobody else was there.

They killed a few chickens and made him dress and cook them. After they ate, one of the soldiers mentioned lying on "that pile of uniforms over there and resting a while" before they moved on. They were following the army to Savannah. Pap told them, "I wouldn't do dat if I wuz y'all." They wanted to know why not. He told them, "Dem soldiers out yonder in dem graves died wid smallpox, and dey wo' som o' dem unifo'ms."

They sort of roughed him up for not telling them about that at first. He told them, "Suh, y'all jest as'd me if anybody else wuz here." They didn't like it, but they grabbed up their guns and bags and soon scooted out from there.

Sherman's men would have killed the captain if they had found him because he was a rebel soldier. After the mop-up crew left, according to Pap, he got the captain out and found him almost dead. He got a quart of whiskey that the Union soldiers *didn't* find and filled the captain up with it to stimulate him.

The next day Pap doped him with black coffee, fed him some soup, and nourished him back to good health. As soon as he was able, they left the camp and hiked up to Chattanooga, where a group of rebel soldiers were stationed. The captain gave him seventy-five dollars for saving his life. That was rich for a Negro in those days.

Now when you read in the history book about the Battle of Chickamauga, you'll see where a Confederate officer was captured by Union soldiers, but he escaped and went to Chattanooga. That was the one Pap Dean saved. Going between Savannah and Chattanooga by Highway 29, there are still some old cannons from that battle on display.

When Pap came back from the war, he lived in Lone Oak part of the time, and part of the time he lived in St. Marks with his grandson Ivy[3] Dean. Ivy's daddy was Jim Dean, who had a brother named Bob Dean. These were called my cousins, but

I do not know how. I know Pap was not the father of my daddy's daddy, so he may have been the father of my daddy's mother. I didn't know her folk.

He had several wives[4] but never lived with a wife that I knew anything about, so I never knew which one was supposed to be my great-grandmother. There was some slave trading before the war, so we cannot "mess with" that.[5] There was a whole lot of "half" and "step" and all in there. I can't explain it all.

Pap had a service revolver he brought home from the war. I didn't see his gun when he was around. He initially gave it to his grandson, at least he claimed him as his grandson, my daddy's brother, Uncle Jim. Uncle Jim gave it to my brother, William, and he gave it to me. That's how I got it. That is why I have the pistol now that Bill had. It's a .38-caliber revolver manufactured about 1840 by *Iver Johnson*[6] and Company. It's on the list to give to one of my boys when I'm gone.

I used to go fishing with Pap. He was a retired farmer. The only work to retire from was farmwork. They didn't call it retire. They were just old and set out on the shelf. We went fishing on what was known as Turkey Creek that ran into Flat Creek, in Meriwether County, out from Lone Oak. Pap liked to go fishing on Turkey Creek behind our house, in a place called Aunt Deena's Fishing Hole.

They called it that because an old Black woman named Deena went there every day. She was about the age of my daddy's mother and daddy. It wasn't running over with fish. People could sit for hours and not catch anything. That's why they used to carry their dinner. Sometime I caught a fish. We fished for minnows, now called shiners, and catfish, perch, and trout. I was about ten or twelve years old at the time. A few years later he died.

Pap belonged to St. Paul CME Church and was buried in St. Paul's Cemetery, the old part. Nobody knows where his grave is now because there was never a headstone. It was just marked by some rocks. It was about 1926. They said he was 113 years old.[7]

Chapter 2

MY GRANDPARENTS

My paternal grandfather was named John Irvin Rosser. He had five daughters. The oldest was Leila. She was my favorite aunt. When I was a little boy, she used to call me Pretty. His other four daughters were Lunia[8] called Sis, Carey,[9] Lou, and Ida. He had three sons: Waymon, Jim,[10] and Robert Miles called Bob. Bob was my daddy.

My grandfather was a sharecropper. The boss on whose land he was living and working was named Bill Lee. When Irvin got old, Bill Lee told him that he would have to move off his place because he was going to have to get someone to stay there who could work the land. My grandfather was too old, and he cried about it. I remember that.

He didn't know how to do anything but farm. Of course, he had been providing for his family ever since he had a family. If he couldn't work, he'd hire out one of his sons. At this time his sons had been marrying and going to other farms to sharecrop, and there was no one left. My daddy was hired out to the Latimer farm.

My paternal grandmother was Elmire Smith Rosser. I did not know her well even though she was my daddy's mother and my grandfather's wife. She was a real big, fair-skinned woman with a pretty face. She had a sister and a sister-in-law that never married nor had any children, so they lived with her and Grandpa Irvin. She and her sister were not dark, but they weren't light as my momma was. That's why I say they were brown skinned.

Her sister was named Jane Smith. We called her Grandma Jane, but that doesn't mean she was a grandma to anybody; it just means she was about the same age as the other women there who were real grandmothers.

Grandma Elmire's sister-in-law, Grandpa Irvin's sister, was named Lou Rosser. Aunt Lou was short, fat, and round, and her toes would not reach the ground. When she was walking downhill, she would get to walking fast, almost running, and she would fall down, *ka-blim!*

My daddy's parents lived up a hill from us on the way to Zion Hill. The years we lived on what was called the Cochran place, I was up there at their house all the time.

One time in a situation me and Margaret, Aunt Leila's little bastard, were playing, and I wasn't thinking about sex or nothing. Grandma Elmire came to our house and told Momma we were "flacking" or something like that. Then, Momma didn't let me go up to Grandma's house. A year or two later, she wanted to send me. I said I didn't want to go, and she didn't make me go. They lived there until they died. That belongs to White folk now.

My maternal grandfather was White. His name was John Barrow.[11] He was the second richest man in that area. Mr. Barrow had two White children by his wife,[12] a girl named Betty Zora and a boy named Jim. They were about the same age as Momma and her brother.

I remember seeing a hundred bales of cotton under my grandfather's shed. He kept that cotton where everybody could pass by and see it. I remember hearing folk talking about it. They wondered why he wouldn't sell some of it. At that time it was going for forty cents a pound on the market. He said, "When it goes to fifty cents, I'll sell it." When the price was going down, he still held out. Things didn't go his way. It never did go up, and he never did sell it. His son Jim ended up taking thirty cents a pound for that cotton.

We didn't receive any of the wealth. My "ole man" was a mite jealous of his White father-in-law, to the extent that he wouldn't accept the fifty acres of land and a mule, which his wife's daddy had offered him. Her brother John Henry Ellis[13] accepted it, but my daddy didn't. So we were sharecroppers all of my life there. John Barrow was not accepted by my dad because he was a White man. That's why he refused to accept the land and the mule he was offered. He didn't want that land because it was off from the road, like[14] they couldn't have built a road to it! He never said a word to me about it.

My maternal grandmother was John Barrow's mistress. That's how come[15] Momma and Uncle John Henry were light skinned. They were half-White. Everybody knew it, and John Barrow owned up to it. She was Jennie Hutchin, born to former slaves.

It was dirty back then, how those slave owners did their slaves. He paid ole John Clark to marry my mother's mother. Gave him forty acres and a mule. But her husband was not the father of her first two children. Later she was married to Bill Ellis. She had another child, Rufus Ellis,[16] and Bill Ellis was supposed to be his daddy.

Grandma Jennie was a good cook, and I always liked to eat at her house. When she cooked fried chicken, I'd get a thigh or something. At home all I got of fried chicken was a wing. At her house I'd get a whole piece of meat. At my house it was so many of us they had to cut meat in pieces.

She lived up at Macedonia. My wife's uncle Newt[17] boarded with her for eighteen years while he was pastor of the Macedonia Baptist Church.

She used to tear my back end up. Pa's mother, Elmire, would hit lightly on the back. I would laugh when she hit me, but Grandma Jennie would beat me. She would hurt when she whipped. She would use anything she could get her hands on, I believe. Mostly it was a switch out of the broom made for sweeping the yard. It was usually made out of dogwood limbs.

We would go to visit three, four, five times a year. Maybe I'd break a glass or something or I'd hit one of the girls, my sisters, and they cried and she would whip me. Once my sisters, Mae and Mire, and I went up there and helped them pick cotton. When we were leaving, she said, "Tell your momma Bro is so bad she ought to cut off his ass, put it in a sack, and send it here for me to beat." She laughed. We all laughed at that.

The Hutchins relatives were a lively bunch. Grandma Jennie had three brothers. I remember Uncle Josiah[18] Hutchin because he was at our house a lot. When I was four or five years old, he told me, "Say, boy, show me your peter. If I like it, I'll give you a nickel." He didn't like it, and I got mad with him.

He was a rounder. He was such a woman's man. He used to carry a bottle of Vaseline in his pocket. He would be visiting all the women and said some of them were tight; that's why he kept that with him. He ran at all the nasty women. One of them was a woman named Ann Holmes. He'd go by her house before going home. He'd leave there smelling like ole Ann Holmes. When Uncle Josiah Hutchin got home, his wife would make him take a bath before he could be with her.

Ann Holmes had a husband, but I guess she was sort of a prostitute. I guess her daughter was too. I remember her daughter wasn't married, but she had a baby in the woods. She started home and fell out. They found her and carried her home to take care of her. When she came to, she told them she had a baby out in the woods, and they went out there and found it. That baby lived.

Another brother was named Warren Hutchin. I must have called him Uncle Warren Hutchin, but I don't remember anything about him. I remember their brother that was supposed to be crazy, or was a terrible fellow, one. He was Uncle Andy[19] Hutchin. He was the fellow that went to Atlanta one time and saw some men had red shoestrings in their shoes. He liked that and bought him some red strings for his shoes.

One day a fellow ran up to him and pulled him around the corner and asked him, "Are you a picker or a toter?" He said, "I'm a toter," even though he didn't know what that was. The fellow stuffed a lot of money in his pocket. When he realized he had the take from a pickpocket ring, he got scared. He caught the train and came home. I don't think he ever went back to Atlanta.

I didn't know anything about Grandma Jennie's sister Julia, who married Bob Thrash. But I knew some of their children. They were my momma's cousins, and we stayed pretty close with them. Loftin Thrash, we called Lost.[20] Lost Thrash came to my wedding. He had a car, and he brought my daddy. Those were the only two of my folk that came.

Lost Thrash lived in Newnan. Once Momma had to have an operation. She had appendicitis and it had to come out. A White doctor did the operation. She couldn't go to the hospital. They didn't take Negroes as patients, so she had the operation at cousin Lost's house.

His wife, Laura, was a big fat woman. One time somebody said to her, "I wish you'd give me some of that fat." She told them, "I would if you could make it stick." It's funny the things you remember about some people.

CHAPTER 3

MY PARENTS

My father did not go to school but twelve days in his entire life. So he didn't know how to be interested in education. He wanted us to learn how to work with our hands. That was his aim and ambition. What he did was not based on education whatsoever.

My daddy sharecropped for Mr. Gene Latimer. He was a farmer. He would give my daddy shotgun shells. He wanted all the quail Pa could kill. My daddy needed the shells, so he'd take ole man Latimer quails every time he found some. He set up close to the house, and when the wind blew, a rush of quails would fly up, and he would pick 'em off one by one, *blam, blam, blam!*

He could kill anything he aimed at. On one hunt he killed eight rabbits. He took me with him. The place where Rob Mays and Wilbert Mays lived had lots of rabbits. There was a pasture that ran by the spring. That water brought the rabbits out. If you stood still, you'd see one or two come tipping down to Little Yellow Jacket Creek. That time I ran the rabbits out. They ran down the ditch and got killed.

He liked baseball. He was a catcher and played on the Lone Oak team. Ours was the biggest community to draw from. There were teams in Hogansville, St. Marks, and Grantville that we would play.

Pa liked the Southern League and kept up with them through the newspaper. It was White teams, but each one of those cities had a Black team with the same name. There were the Chattanooga Lookouts, the Memphis Chicks, the Nashville Volunteers called the Vols, the Atlanta Crackers, the Birmingham Barons, and there was a team in Little Rock. They tried to get a team in Florida, but somehow it didn't ever happen. I think the people thought they were better than Georgians and Alabamians. So they never joined the league.

Neither one of his brothers played baseball. Uncle Jim played the guitar, and Uncle Waymon played the bass viol. It was almost taller than he was. But when he pulled that bow across it, that low *va-woooom* sounded good. They just played at home. There weren't any instruments in church. In church everybody got the tune by singing shaped notes. Every tone on the scale—do, re, mi, fa, sow[21], la, ti—had a different shape on the music staff. They would sing a song by the sow-fa syllables to get the tune then sing the words to the hymn.

My daddy was a great note singer, and my momma too. I used to be able to sing notes, my wife could too. But we lost it. If you don't need to use a thing, it will eventually leave you.

My mother was named Caroline[22] Ellis. She was very light skinned. She had long hair. She had two plaits of hair that would hang down in her lap when she would sit down. She used to like to eat peanuts. I remember seeing her in a chair with parched peanuts in her apron. She would have to shake her plaits to get those husks out of her hair. She was a very beautiful person with a quiet, loving manner.

Momma and Daddy were married[23] early in the 1900s. She bore eleven children. One child, Leila Beatrice, died in infancy in 1905. Momma was very intelligent. She wrote a poem dedicated to Leila Bea's memory. She taught it to the rest of us. I still remember it:

> We are only five little brothers
> And five little sisters, dear.
> One little sister's gone away,
> We are sorry she's not here.
>
> We loved her, yes we loved her,
> But, angels loved her more.
> And they have sweetly called her
> To yonder's shining shore.
>
> A golden gate was open.
> A gentle voice said, "Come."
> And with farewell unspoken
> He calmly called her home.

Here's another one she taught us. She could have learned it in school, but I believe she wrote it:

> A baby came to our house
> Not very long ago,
> And it seemed to me like mother
> Wouldn't love me anymore.
>
> But she took me in her arms
> Just like she used to do,
> And told me that a mother's heart
> Was big enough for two!"

Black women had to depend on midwives. But because her daddy was John Barrow, she could be attended by Dr. Latimer. That was the White doctor in town. The Latimer place was back in the woods, but he and his brother Gene moved to town. One time Momma was having a baby. She thought it was coming soon as the pain started. It didn't. He went to sleep. She had to wake him up when it was coming. He said she was lacerated, and he sewed her up.

Momma was a good cook. She could cook some chicken dressing that would make you bite your tongue. I think my favorite meal was roasting ears, corn cut off the cob and fried with fatback, and buttered biscuits. I also liked sweet potato pie. Sometime the pastor would eat dinner at our house. I wasn't allowed to do anything but stand and look. I couldn't go to the table until he was through, but he didn't eat up all the chicken.

She took good care of us. She used to piece quilts and patch britches and shirts for us. I don't ever remember her having a job away from home. When we got to farming, she would bring us some water to drink. It tasted better than any other water. She would sometime hoe cotton in the morning, then go home to cook. Bless her heart!

She completed the seventh grade in school, which was all that was available for Blacks in Lone Oak. She taught my daddy to read. He subscribed to the *Atlanta Constitution* and read it every day until he had a stroke.

My maternal grandfather's influence went a long way in determining how we fared. My mother and uncle both were very light skinned. Their mother, who was a slave, was kind of brown skinned. It was a race situation. They couldn't really function as a person in the community because of their color, so when they were grown, they married darker people. My mother married my daddy, who was dark skinned, and Uncle John Henry married a woman named Victoria Cousin,[24] who was very dark. And when she died, he married Lilly Holloway, who was also dark.

My uncle and my daddy were churchmen. They both were among the men that helped keep St. Paul Colored Methodist Episcopal Church[25] going. My uncle, according to his rating, was one in the higher echelons of the church because he had his own land. The way a person gets along financially has a lot to do with his christian walk.

There was competition between his children and us because they had all that they wanted, and we were living from hand to mouth. My daddy didn't have his own land, so we had to come up in poverty, sharecropping.

This was like most ex-slaves and their descendants during my time. That was about a hundred years ago when all of this happened because I'm ninety-five years old. When you had your own land, you could grow cotton. "Cotton was king!" The only way any real money was made by Blacks was with cotton, and men who worked on the railroad.

Father, Robert Miles "Bob" Rosser, 1960

Mother, Caroline Ellis Rosser, age 18

Endnotes, Lone Oak, Georgia, Circa 1813-1906

1. "The personal slave that was given to White children usually helped them until they reached adulthood. The richer males from the south bought their own uniforms, their own guns, and brought their own personal slaves when they went off to war" (Personal communication, M. Suejette Rosser Riley, July 2003).
2. He was a look-alike for JR's brother-in-law, husband of his sister, Mamie Rosser Johnson Shuffer.
3. Ivy Dean's name was pronounced "eye-ree," as if the letter *v* were the letter *r*. It may have been an adopted spelling of the word *ivory* used as a name.
4. According to Freeman and Freeman (1985, p21), Peter Dean's wife, Sinia (sign-nigh), was the mother of Sophia, Bob, and Jim Dean.
5. The term *to mess with*, as used here, meant to conduct an inquiry about a private matter. It's other meanings are "to touch" and "to bother" or agitate.
6. Iver Johnson Gun & Cycle Works, Fitchburg, Massachusetts.
7. "People who lived to get old back then looked older than they were. Pap may only have been ninety or so. It was still a long time to live" (Personal communication, Minnie Freeman Ellis, daughter of John Henry Ellis, July 2003).
8. The name Lunia was pronounced "loo-knee."
9. The name Cary was pronounced "kay-ree."
10. Jim was likely a nickname for James, but this was not verified by a birth record.
11. John Barrow's death certificate from Meriwether County Health Department, Greenville, Georgia, confirmed dates of birth and death, also place of death and burial as Lone Oak, Georgia.
12. Celestia R. Sewell Barrow (March 21, 1857-December 8, 1931), wife of John Barrow, mother of James H. Barrow (November 2, 1882-February 16, 1950) and Betty Zora Barrow (September 7, 1899-July 16, 1945), Allen Lee Memorial United Methodist Church Cemetery, founded AD 1844.
13. Uncle, John Henry Ellis, January 14, 1880-December 12, 1960.
14. The word *like* is used to mean "as if."
15. *How come* is a term meaning "why" and may be a statement or an inquiry.
16. Uncle, Rufus Ellis, May 3, 1892-October 3, 1931.
17. Rev. Newton Moreland was married to Estella Jane Lynch, sister of Samuel Elisha Lynch, who was the father of Pauline Lynch Rosser, wife of JR Rosser, Sr.
18. JR pronounced *Josiah* as "joe-sigh."
19. Antina (Freeman & Freeman, p1) pronounced "an-teh-ney" by JR.
20. The nickname may have originally been Loft that evolved over time to become Lost or Loss.
21. "Sow" was the pronunciation for sol, Italian pronunciation "so," "sew," which is the fifth note of the scale. Blacks used the sol-fa syllables, or solfeggio, in transition from singing by ear to reading notes on the musical staff, hence the term *note singing*.

22. She was likely named after her mother's mother, Caroline Hutchin (Freeman & Freeman, p1).
23. Caroline Ellis and Robert Miles Rosser, married October 24, 1901
24. Aunt, Victoria Cousin Ellis, October 4, 1880-August 23, 1926
25. The name was officially changed from *Colored* to *Christian* Methodist Episcopal Church on January 3, 1956 (Encarta, 2000).

MERIWETHER COUNTY

Year	Age	Grade	Events	
1906			Born December 30	
1907	1			
1908	2			
1909	3			
1910	4			
1911	5			
1912	6			
1913	7	1	Entered school	School year (SY) 1913-1914
1914	8	2	SY 1914-1915	WWI begins
1915	9	3	SY 1915-1916	
1916	10	4	SY 1916-1917	AURS organized
1917	11	5	SY 1917-1918	
1918	12	6	SY 1918-1919	Plowing one-half year WWI ends
1919	13	0	SY 1919-1920	Plowing
1920	14	0	SY 1920-1921	Plowing
1921	15	0	SY 1921-1922	Plowing
1922	16	0	SY 1922-1923	Plowing
1923	17	6	SY 1923-1924	Plowing one-half year, enters boarding school

Chapter 4

EARLY YEARS

I was born December 30, 1906. My momma named me John Robert Rosser. I was named *John* after my grandpa, John Barrow, and her brother John Henry Ellis; and of course, *Robert* was after her husband.

There are a bunch of *Johns* down there in the Rosser situation. When it comes down to what happened before the Civil War, looking at the gravestones in St. Paul's Cemetery, you don't know whether you are looking at your own grandpa's headstone or not.

Momma started calling me Brother because I was the only brother for a while. Then they started calling me Bro.[1] So that was my nickname that stuck. My family called me that as long as they lived. Some of their children still call me Uncle Bro. Some call me Uncle John Robert.

We lived in a big five—or six-room house. It was on the road between Lone Oak and Grantville. Next to the back porch at one end was my room. That part of the house went out to the well. My room was two or three feet from the well. The window didn't have glass panes. It had nothing but a shutter. When company would come, they would pick up the dirty clothes and throw them in my room and shut the door. The two girls' room was where company would often go. They had curtains. I didn't. Momma and Daddy's room was up-front with a fireplace.

I remember the first automobile that came up that road when I was about five. I saw it! Where we lived was sort of a long hill. There was a mud hole in the road, out in front of our house. Wagons had made the mudhole after a rain. The car ran up in that hole. It was a T-Model Ford.[2] It got to going loud and snorting around and got stuck there for a while. When it got out, I was in the house under the bed. I was afraid. It was the first time I'd seen an automobile. I didn't know what it was going to do. I was scared of it.

My daddy never owned a car. He drove a dray. He had to go to town every morning and get what would come to Tom Turner's store. His son Moses and my daddy were friends. Pa worked a year for them for wages before he got married. He continued to do that for a while. He would go to Grantville and get the freight and bring it back to the store. He stopped driving the dray and started sharecropping. That was the only work that I knew he did. I was four years old. I liked picking up potatoes.

My brothers and sisters[3] and me in the order we were born are Jennie Mae, the oldest daughter, and Elmira, the second daughter. Leila B, the third daughter, lived only a few months. Then I was the first son. Mamie Frizzell was the fourth daughter. William Henry was the second son. Amy Brown was the fifth daughter. Joe Nathan, Ellis Holsey, and John Irvin were the last three boys, and Laura "Bob" was the baby. That made eleven of us. Ten of us lived to get grown, five girls and five boys.

Momma taught us to pray. She taught her children this prayer, and I taught it to mine: "Now I lay me down to sleep, I pray the Lord my soul to keep. If I should die before I wake, I pray the Lord my soul to take." She taught me to add this part: "God bless Momma and Daddy, bless my brothers and sisters, and make me a good boy. Amen."

When we were little, I do feel like Momma and Daddy treated us all the same. Not alike, but all the same. If something happened, they would be sort of nice to you. After you got better, of course, that vanished. Here is something I remember about each of my brothers and sisters.

Jennie Mae. Momma was sickly and couldn't necessarily spank us as we should be spanked. I did something I shouldn't have done. I threw some trash in the pot where my sister was washing. She told Momma. Momma was in the house in the bed, sick. She said, "You just have to whip him, Mae. I just can't do it." Mae whipped me, and I thought that was terrible. But I respected her more after then.

Elmira. Well, Elmira was my buddy. Everything they asked me why I did it, I told them, "Mire told me to do it." Sometime it would be the truth and sometime it wouldn't. But I had to have an alibi, and that's what I used. I remember that well. I remember Momma saying, "Bro, everything he does, good or bad, Mire told him to do it, if you asked him why he did it."

Mamie. She was a little spitfire. We would do concerts together. I remember we had a little play and we'd sing "Rock it like I used to, rock it like I used to, rock it like I used to, and I ain't gonna rock no more, oo-oo!" We'd twist, then bump. You know, put our butts together and bump. So she was my pal in such things as that, in plays and things. I don't know what you call them now. Anyway, she was my "bumper." I could put a ring in her nose and lead her anywhere, that is, I could, as you might say, suggest anything to her.

Amy. She was nice. She was a good pal till you made her mad. Back then, we were learning bad names to call each other. When I made her mad, she would call me "ole doggone, dad-burn devil." I remember she was bad about saying that. See, we weren't allowed to use *devil* and *dad-burn*, and *doggone*. We weren't allowed to use those expressions. She would use them whenever I'd make her mad. Maybe hit her or something like that. But she was a buddy of mine.

William. I played with him. I had a little brother to play with, but I played too rough, often making him cry. And when he'd cry, I'd get a whipping about it. I remember we were over in the field picking peas, and I ran across a tortoise, a terrapin—we used to call it a "t'apin." I picked him up and showed him to William. He asked me what it was, and I told him it was a "t'apin."

I threw the terrapin toward him, didn't mean to hit him. It hit him on the head, and he started crying. The shell was hard, but it hit William's head sort of on a glance. And it didn't hurt him. He didn't bleed or anything like that. I got him stopped crying before we got home. I didn't get a whipping about doing that, but I got a whipping about several things like that.

We didn't eat terrapins. We teased them. They would shut up in a shell, and you couldn't see their head. We learned how to put fire to their tail and make them stretch out their head and front feet and crawl. That's the only way we could get them to crawl. Lord, have mercy!

Joe. Joe was kicked in the eye by a mule. We were testing him to see if that eye was put out. He could see a long way with that eye. He could see a rabbit running on the side of the road. It was a cut above his eye. Pa carried him to the doctor and had him to sew it up. He was sort of petted by the family for a good long while.

He was standing up in the wagon, chunking the mule under her tail with a piece of cane, and she kicked him. Of course, Cora would kick anybody, anything, anytime. She was just a kicking mule. She would be eating and haul away and kick the back end of the stable. After that, Pa drove her to Grantville and beat her all the way there and all the way back. But he couldn't afford to get rid of her.

Mules cost around three hundred dollars in that part of the country. That was a lot of money. Many a poor Black man has worked and given his hard-earned money for a mule. Then messed around and the mule got sick or died or something like that. That was three hundred dollars lost. If a mule could not plow, he wasn't worth a shit.

Ellis. When he was little, he was jealous of the whole rest of us. I don't remember whether he'd cry if my daddy had a bigger piece of meat than he did, but I know if any of us did, he'd have a fit. Momma handled it. At that time he had that ole carbuncle on the side of his neck, and it bent his head over. It was out there like that. He cried for three days and three nights just steady along. And when that thing came to a head and burst, he went to sleep while it was running. It'll quit hurting, you know, when it bursts, and Momma had to clean things up.

When he was a man, I remember a true story he told us. He was in the army during WWI. He was sent to Okinawa with General MacArthur. One time Ellis got drunk, and the general busted him down to private. Later on, the general was in the jungle and a big snake, about twenty feet long, swung his head down out of a tree and was staring him in the face. Ellis was a good shot and killed the snake. The general gave him his rank back.

JI. He was named John Irvin after our Rosser grandfather. He always had some money. If you needed a dime to get some smoking tobacco or something like that, and he was around, you could borrow it from JI. I asked him, "How come everybody wants to borrow money from you?" He said, "Ah, 'cause I'm rich!" He always kept some money, and all of us borrowed money from him, even up to Pa.

Laura. I don't remember anything about Laura as a sister there at home. I don't even remember how she greeted me or how I greeted her. I just don't remember.

It's sad to say, but I know it's true. My daddy had other children. Back then, the men of the church would visit widows. They said they were "taking care of them" like it said they were supposed to do in the Bible.[4]

They would go and cut wood and draw water and so forth. Some months later, these widows would have a new baby. Those men believed they were doing the right thing. They stepped out of bounds in those situations. They weren't too far removed from the jaws of slavery.

I had a half sister named Lulene Houston. She married Henry Louis Freeman. They had ten children. She sent her children to college. Pa fathered Lulene while he was "helping out" the widow Mattie Cousin Houston. She was a sister of Victoria Cousin, who married Uncle John Henry. After ole man Houston died, his widow had more than one baby. They had different daddies. They were different colors.

For a number of years Uncle John Henry "took care" of a widow named Elizabeth Phillips. He fathered two[5] babies. When their mother passed, her oldest son carried those two, knocked on Uncle John Henry's door, and handed them to him.

When I was a little boy, I got a whipping every day. When Pa would get home from work, he'd tell me to go get a hickory switch. Then he'd whip me. I learned to holler and cry real loud so he'd slack up on the licks and I'd stop crying.

Sometime I didn't know what the whipping was for, and sometime I don't think I deserved it, but he whipped me anyhow. He told me every lick he gave me was one less a prison guard would give me. I guess he was right because I have never been arrested and never gone to jail nor prison.

Once Mae and Mire got some apples out of his trunk. He had been saving up the apples for Christmas, but he had his money in the trunk as well. I got a whipping about it even though I didn't eat any of the apples. We looked forward to Christmas, although we didn't get anything but an apple and an orange and some fancy candy. It was not regular stick candy; sometime it was coconut or caramel.

I liked the fruit of the woods. Hickory nuts, chestnuts, chinquapins, black halls, and of course apples, peaches, and plums. I loved black walnuts. We had a big black walnut tree. It had so many walnuts even in the summer they would still be on the ground. They didn't rot.

There was a little hill between our house and a family of Smiths. I remember Mr. Smith making a fire outside and letting me warm by it. He was Bennie Lou Smith's daddy. She married Calvin Ellis. I was five or six years old. Mr. Smith gave me a little air rifle. I was excited about it. Pa told me to take it back. I didn't want to and I buried it. I had nightmares. I thought I heard the door rattle. I was scared. Sooner or later, that feeling passed. That rifle never turned up.

I remember Pa helped to buy a wedding ring or wedding dress for somebody in Mollie Smith's family. He wasn't related to any Smiths. He and Momma had some stiff discussions about it. One time after the argument, I saw him do something I didn't like. He put her across his lap and pulled up her dress and spanked her. That was wrong. He shouldn't have done that. She didn't deserve it.

One thing I never understood was why Pa called me a "sharp-headed devil." I don't know why he did that. He didn't call any of my other brothers and sisters that. It hurt me because I always did what he asked me to. I was always humble. I had to do what he and Momma told me to do. I was afraid not to. I didn't think I had any rights.

Momma knew the best thing to do in most any situation. We had a dog named Kroft. And it seemed he knew all of us. When a child would be born in our family, when conditions presented themselves, she would call him in and let him smell the baby's hair. From then on, he knew the baby.

Chapter 5

SCHOOL AND CHURCH

When it was time for me to start school, we didn't have money to buy me a first-grade reader. Back then textbooks weren't free. One of my sisters had a second-grade reader. Pa asked the teacher if I could use it until he could make his crops and he'd buy me the first reader. She told him yes, so that was the book I carried to school. By the time his crops came in, she told him I didn't need the first reader because I was doing just fine with the second reader.

I went to a school that offered four months in the winter and two months in the summer. You've heard other folk say they had to walk two miles to get to school. It was true. I'm not just repeating it. It was about two miles from where I lived, and I walked. Everybody walked. Even the White folk in Meriwether County didn't have a bus to ride to school.

I liked school, for the most part. If I was in school, I didn't have to work around the farm. I remember one time the teacher tried to comb my hair. She made me get down on my knees in front of her, with my head in her lap. She started combing, and I started crying, turning my head from side to side. My nose was running, and I got snot all over her dress. It was one of those shiny black dresses, and you could see the snot trail everywhere. She finally gave up and called me a nasty little boy.

One year I had to pass my teacher's house. Her name was Mary Step, and she was my favorite teacher in elementary school. My next favorite was my cousin Eudora Booker.[6] She promoted me to the fourth grade when I did a problem in long division. Back then teachers promoted you when they decided you were ready. When you could spell days of the week, cities, and things like that. The years didn't matter.

There was an old empty house that had an old piano in it; sometime I would go in there and "bink" on it. But I liked to play games with the other boys, my peers. We'd shoot marbles or play baseball. Baseball was really the only sport we knew anything about and the only sport we had the tools to play. Pa would make us a bat. I could catch balls hit high. When the big boys and men would be playing and they'd hit a foul ball, I would catch it and throw it back. I didn't get a chance to bat, but I could catch, pitch, and bat.

One game we played with a ball, and bat was Mollie Bright. The first person would say, "Molly, Molly Bright," and pitch you the ball. The second person would hit the ball and say, "Sycamore and ten!" Then run to the base. If they got the ball you hit before you got to the base, you were out.

Another game we liked to play was "cat-ball." You had four players: a pitcher, a catcher, a batter, and a fielder.[7] You'd hit the ball and run. When the play was over, we'd all change positions. There was a level place down below the schoolhouse, about as far as from here to Elrod Street.[8]

We'd go down there to play ball at recess time. One time we didn't hear the bell ring and looked up to see the last child going in. We ran as fast as we could, but the door was closed by the time we got there. The teacher got a switch to whip me, and I didn't think that was right. So I snatched it from her. I broke it and threw the tree limb in the kindling box. Of course, I got a beating from Pa when I got home.

Sometime we'd have a little fight walking to school and back home. We made little ole whistles out of reeds. Somebody would borrow your whistle and blow it till it wouldn't blow anymore; then we'd fight about it. We didn't tell the teacher about our little fights. The older fellows took care of one little group and made them follow the rules when we would fight.

When they had to stop school and plow and do other work on the farm, I caught it. One day when school was over, I ran on ahead. I gathered up a pocketful of rocks and climbed a tree where I could see them and they couldn't see me. When those bullies walked by, I pelted them with rocks. *Ping! Pow! Pow!* They went to grabbing their heads and hollering and running. I don't remember starting one of those fights. I wasn't a bully. But there wasn't anybody going to bully me. I wasn't scared of nobody but God and nothing but lightning!

Unless it rained, Sunday was the only day we didn't plow or work in the field. We went to church. The only church I ever joined in my whole life was St. Paul CME Church in Lone Oak. The week I joined, it was a revival, and we had meeting[9] every night. There was no electricity in the church. It had kerosene lamps sitting on little shelves. A lot of my peers joined that same week. At age nine, I thought we had the devil on the run!

Elder Dudley baptized me. He poured some water on my head and wiped it off with a towel. That was the way Methodist baptized. They still do that. The pastor was Rev. G. S. Myrick. He told Momma that I was called to preach and just didn't know it.

When I heard that, I went out in the field and tried to preach. I hollered loud until I got tired. I said anything I could remember hearing a preacher say. I practiced so hard my throat got sore. I stopped trying. I felt that maybe God had a "special calling" on my life, but I never did really understand what it was. I was a little disturbed because I was on the fence with this religion thing. Then after I tried to preach and that didn't work, I didn't really ever think of how it was influencing my life.

You didn't take communion until you joined the church. We had it every first Sunday. It was just something you could do. And some folk couldn't do it if they weren't members of the church. The Baptist folk wouldn't give it to the Methodist. They thought it was something unclean because you drank out of a cup then gave it to the next person and they drank. But some Baptist did that too.

I remember some of the big fellows would like the girls and walk them to the church. Once it was raining, and I said, "You want it to stop raining?" They said, "Yeah." I jumped up and patted my hands and did it again. It really did stop raining. For a while I thought I had something to do with that. But it was going to stop raining anyway.

We lived in a mean world. I remember that White folk turned out their church before us. Some of those little White boys would get down to the stream crossing before we would. It had a catwalk that was just two planks wide. Two or three of those boys would stand on the path blocking the way and you couldn't pass. They would laugh when we had to get off the path and get in the water to go on home.

Them little ole White boys called us "niggers" all the time. You couldn't say anything because you might get killed when they reported it. We would be walking by, and they would say, "My daddy'll kill you niggers and throw you in the Flint River." That's what was supposed to have happened to a lot of Black men. Black women have always been pretty. White men would see them and want them and have a way of getting them. When a Black man objected to that kind of doings, why, they would kill him. There wasn't any law against killing a slave.

All the water around there—Big Yellow Jacket, Little Yellow Jacket, and Turkey Creek—ran into the Flint River. Where I lived was the bottom end of Appalachia, the fall line. When it rained, water in those streams would move fast, but the farther down the fall line it went, the slower it got. It carried mud, gold, salt, and all such along; and by the time it got down there to the Flint, the water was pretty muddy.

Chapter 6

GROWING UP

As a boy I was always getting stung by something. Those yellow jacket stings weren't much, but those red wasps were tough, worse than the black wasps. They would hurt. I used to see a wasp nest, a hornet nest, or a beehive and throw rocks at it to knock it down. When it hit, that movement would shake them up, and they would come swarming out.

They always knew right where you were and would sting you before you could run and hide. They knew if they hit you where you had clothes on, it wouldn't get you, so they came right for your face. There would be so many stings around my eyes they would be swollen shut the next morning.

I rode the mules, cows, pigs, just about anything that had four legs. Once Pa got a new bull. Every time he went to town, he'd tell me not to ride that bull. Well, one day he forgot to tell me. That was the day I decided to ride that bull. I lured him over to the fence and jumped on his back. He took off and I hung on. I rode him till I wore him out.

You know, it's a wonder that boys live to get grown. We do so many foolish things. I should have been dead. I got this scar on my forehead when I was about nine years old. William and I were playing "king of the hill." He was on the porch, and I was trying to get up on it using a post. It gave way, and the roof fell down on my head. I've had this scar ever since.

There was a hedgerow between the Latimer farm we were working and Uncle John Henry's place. I went in that hedgerow because I saw some "bullets"—that's what we called muscadines. I didn't have on any shoes. I was looking up and waving through the hedge, thick with briars, and felt a keen cut on the side of my foot. I thought it was a locust bush because they have a limb with stickers and stiff thorns.

I was bitten by a rattlesnake. I saw the rattlesnake had drawn back and was fixing to strike again. I got a big rock and killed it. It was a big snake about three feet long and about an inch and a half or two thick. It was a pilate—that's a female snake. I know because its belly was smooth and yellowish. The male's is rough and brown.

Rattlesnakes won't run from people like some other snakes will. They'll fight you. But folk say if you don't touch them, they won't touch you. Back then when a snake bit somebody, they died. I thought I was going to die. To tell you the truth, I have been scared of dying all my life.

William was there, but he couldn't keep up with me. I ran all the way home. It was a half a mile to a mile. When I got home, blood was running out and streaked all across my leg. My foot was swollen up to twice its size. Pa wasn't there. Momma tried to wrap it with something, and I got in the buggy and drove to town, to Grantville, to Dr. Culbert.[10]

He owned the drugstore and was a pharmacist, not a regular doctor. When WSB radio station in Atlanta had just formed, Dr. Culbert took an ad out on the station. I remember what it said: "Ask Dr. Culbert for a brick, he'll hand you one right quick!" A brick was a block of ice cream.

He poured turpentine on that snakebite. That didn't hurt. He wrapped it in some gauze. He said he wouldn't wrap it tight because it would blister. I worried less about the blister than I did about dying. When I got back home, my foot had gotten even bigger.

I don't know what that turpentine did. I guess I ran most of the poison out, and the turpentine did the rest. The swelling went down, and it got better. I maybe stayed off that foot about two days. While I was "laid up," I remembered the story about the boots that killed three out of four men.

There was a man who had been ditching in a pair of waist-high, government-issue boots. One time after using his boots, his leg swelled up. They didn't know what happened, and he died. They put up his boots. Children were careful about their parents' things. It was a sacred trust.

Time passed and his sons grew up. The oldest brother put on the boots. He felt a little scratch but went on to work. That night his foot swelled up, and he died. They didn't know what was wrong. You can't cure a thing unless you know what it is. Later the next brother wore the boots, and the same thing happened to him.

Finally, the last brother took a notion to examine the boot and found in the back of the boot a snake's fang stuck there. When those men would walk, their leg would press on it; it would stick them, and some of that poison that was in it would go in their leg. Two things folk used to die with, that they stopped dying with, were rattlesnake bites and broken necks.

I was stung by wasps, stung by hornets, and bitten by a rattlesnake. Either one of those could be deadly, but I survived. Those bites left so much venom that nothing else would take on me. I was given the smallpox vaccine three times, and it didn't take. In a few days the skin would be smooth again. When it takes, it leaves a round scar about the size of a dime with little ridge marks on it.

I was still immune to smallpox. Once Fannie Lee, Tom Henry, and I went exploring on the land back of Grantville Road. Ben Booker had smallpox and was isolated over there. We stumbled upon his camp. We found where he had emptied some sardine cans. Those cans were in a sack where the ants couldn't get to it. We sopped them all.

They were all good. After we left, they both had smallpox, but I didn't, and I know I got my part of the stuff.

Ben Booker didn't die of smallpox. He came back into circulation and back to church. He had spots all on his face and on his arms where he had scratched. They say when it begins to get well, it itches so bad you can't keep from scratching it. He lived to be an old man.

When most folk got sick, if they couldn't get a doctor, they just had to die. There was one telephone line at one of the three stores at Lone Oak. If you wanted a doctor out of Grantville or Hogansville, you had to call them from the crossroads. One road ran north and south between Grantville and Greenville, and the other ran east and west from Hogansville to Luthersville. Folk didn't have telephones, nor electricity in their privates homes.

World War I. I remember those times. They had to have somebody like me to keep history going. I just remember they first registered men from about age twenty-one to thirty-one. They killed so many of them they were about to run out of soldiers, and they registered again. This time it was men from age eighteen to forty-five.

My daddy was about forty. He went to register. I tried to register too. I was about twelve, but I looked older than that. The man at the draft board was Atticus Sewell. He said, "John Robert, you not eighteen. The reason I know you're not is my daughter, Martha, was born the same night you were, and she's just about twelve." The war ended before Pa was drafted.

I saw the first airplane about 1918. We heard a lot of roaring where we lived. There was an airfield in Georgia down around Americus. WWI was over. Four airplanes, the double-winged airplanes, came up there, and it was one afternoon when we first saw them.

They flew up high enough for us to see them from where we were. It was the talk of the community because some folk didn't see but three and others saw four. It was really some experience to have seen the "airships," as we called them. We had heard of dirigibles, but none had come through our community, that I knew of. They were supposed to be flying on helium or some gas that was lighter than air that would carry them up in the sky.

Grid Map, Meriwether County, Georgia

North Carolina — Cherokee Indians
NORTH

- ○ Family House
- ❖ Town
- ■ Baptist Church/Community
- ▫ Methodist Church
- ☐ County Seat

☐ Newnan

S P A U L D I N G

▫ Lowell Temple CME
❖ **Grantville**
C O W E T A

■ Macedonia BC

■ Liberty Hill
Linzie Rosser ○Bob Rosser Warner Rosser
❖ **Luthersville**

BIG YELLOW JACKET

○John Henry Ellis
▫ St Paul CME Church
❖ **Hogansville** ○John Barrow

SPRING

❖ **LONE OAK** ■ Zion Hill BC
Irvin Rosser   Walt Rosser

T R O U P

CREEK

John Brown Rosser
Cochran Place
❖ Primrose
Ben Rosser

LITTLE YELLOW JACKET

❖ Saint Marks
William Rosser

Approximate Scale
1 block = .25 miles

☐ **Greenville**

U P S O N

H A R R I S   T A L B O T

Area of Meriwether County, Georgia - Circa 1906
SOUTH
Florida — Seminole Indians

WEST — EAST

Grid Map, Meriwether County, Georgia

Chapter 7

SAME NAME

There were seven sets of Rossers in the area of Meriwether County. None of them were related, at least we didn't ever prove that we were, but we claimed kin. There were different skin colors among them, and now I wonder how they were all named Rosser.[11] As a child, though, I didn't wonder anything about it. I just thought they were kinfolk.

Nobody told us anything back then. Even if grown folk did know, they didn't tell children anything about who was related or how they were related. If you asked, they would tell you "none of your business" or you would get a whipping. If you said anything, you got a whipping.

I remember what happened to one fellow. He said something about his family's business, who somebody's daddy was, or something like that. He was ostracized, called "crazy," and shunned. They believed that "kids," as children were called then, should be "seen and not heard." That might account for so many gaps in my memory.

I knew all those Rosser men and some of their children. They were along about the age of my grandpa, my daddy's daddy, *John Irvin Rosser*. His grave is there in St. Paul's Cemetery.

One of the men was named *William Rosser*. He was the oldest one that I remember. William lived in a place called St. Marks. It was down below Lone Oak, kind of southwest. We all referred to William as Uncle William.[12] Maybe that's why we thought he was a relative. I remember that he was light skinned and owned some land. He had a son named John Henry Rosser, who was known as Peck.

Peck Rosser was dark. He was cousin Beodessa's husband. They were not kin to us, but we always called each other cousins. They had five children: Lillian, Zelda K., Marvin, John Lynn, and Susie. Cousin Beodessa was a schoolteacher. William had two sons by another wife. One was named Marvin; the other was called Sun. That's what they always called him.

Peck Rosser owned a pea thrasher. When people would pick their dried peas, they would bring them to him to thrash. No money was exchanged, but every fourth peck of peas went to him; that's why they called him Peck. People called his first daughter Lillian Peck.

The pea thrasher was on a wagon pulled by two mules. It took two men to grind thrashers, one on either side. One man poured the peas into the hopper. The other man turned the crank, the grinder. The hulls sprayed out the side, and peas came out of the front into a boxlike tray. Peas were a staple crop. Nothing was thrown away. Pea hulls were mixed with salty water to feed cows.

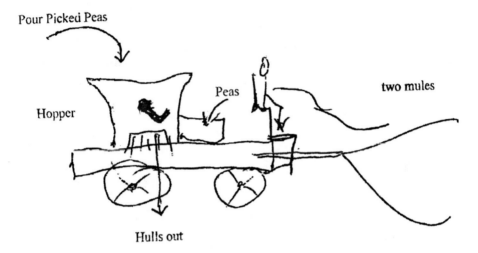

Pea Thrasher: Sketched by Suejette from Dad's description

Pea Thrasher sketched by older daughter in 2003

Ben Rosser was a resident of Primrose. That's still in Meriwether County, but way down in the southeast corner. He was one of the founding members of the American Union Relief Society and was first worthy grand secretary. I inherited his job. Well, I didn't inherit it but finally did end up with his job in the society. He was light skinned. Ben's daddy was a White man who gave him a good start. He got land from his daddy, and he owned an automobile. Those White men claimed the Black children they had by their slaves and gave them land. They did what they wanted to do.

Warner Rosser[13] was up at the Macedonia community, out past Grandma Jennie's house. That was between Grantville in the north and Luthersville on the east. I don't remember anything about him. But he had a daughter, Veronia, who never married. Veronia Rosser taught school. She had four brothers: Arthur, called Coot; Tom; Walt; and Horace.

His son Walt[14] Rosser lived in the Zion Hill community. That was just a little ways east of Lone Oak. That's where the Zion Hill Baptist Church was. In all these communities, there was a Baptist church by the same name as the community. The church was also the first schoolhouse in the little community.

Walt Rosser had children that grew up along with us. He had two daughters and two or three boys along about my age. One daughter, Irene, was a kind of girlfriend of mine. I liked her, but we didn't know whether we were kin or not. My children probably remember Vallee Rosser. She was his daughter. She had the worse case of halitosis I ever saw.

All the men had families. Pa thought because of the last name, we were all kin. He didn't go back into it to really see if it was true. He must have always believed it. When I called myself trying to court some girls that I liked, he told me not to be going with any of the girls and especially not to be thinking about marrying any of them because we were probably kinfolk somewhere down the line.

Another *Rosser* lived in the Liberty Hill community. It was west of Lone Oak. We went through it going to Bessie Ector Rosser's house in Grantville, a little slick place in the road that you turn off on the way from Hogansville, not due west. Grantville was north and Hogansville was west, so it was kind of diagonal between them, kind of northwest. I just can't remember his name.[15] His wife was Anna Rosser. Everybody liked her. She was a midwife.

Three widows lived in a triangle. Aunt Anna lived on the west side of Lone Oak. On the east side was a lady named Aunt Mitt Carter. The other was Aunt "somebody." It could have been Bro Lo's widow. Brother Logan got too old to preach and died. He had a horse called Logan, and everybody called him Bro Lo. I say I remember that, but it happened before I was born. I just remember hearing people talk about him.

I didn't know all Anna Rosser's children, but I knew two sons and two daughters. They called one son Bud and the other one George. Both of them were preachers, and both went to Paine College in Augusta. They were itinerant ministers in different places, but neither of them ever pastored St. Paul. You know how the CME church is. They assign you to different places.

When Paine College was established by some White folk from the north, they put a White man as president, and they wanted to keep it that way. But the first Black president of Paine College was a former classmate of mine from Forsyth by the name of Harold Pitts.

One of Anna Rosser's daughters was named Edna. She married a Reese. People knew Edna Reese. Her particular claim to fame was she cut a new set of teeth at age sixty. She had a plate of false teeth, and those new teeth pushed the plate out, and she couldn't shut her mouth.

Another daughter was named Cora Rosser. She married John Fitten and had two or three boys. She went to Chattanooga. After her husband died, she came back to Lone Oak. This was shortly before Uncle John Henry died. She got in good with him and was connected with the AURS Grand Lodge for a while because I had to call her "Cousin Cora Fitten" before I was grand secretary. That was when the lodge had bloomed out to its highest number of lodges and members. That number was declining when I became worthy grand secretary.

There was one other Black family of Rossers in Lone Oak. They lived in the next house down the hill from us when we lived up on the Cochran place, down the little hill about a quarter of a mile and across a little creek. A little piece up from their house, the road curved on to Zion Hill. They were not kinfolk for sure.

We always called him *Mr. John Brown Rosser*. He and Pa were about the same age, but we were a little older than they were. His oldest child was my age. His wife[16] was the one who raised so much sand about the pig.

Our dog, Kroft, would get anything we told him to get and work on it until we told him to stop. Their pig was in our corn patch, so I told Kroft to "get him," and Kroft went to biting him on the leg and chased him back to the porch of their house.

The pig's leg was bleeding. Mr. John Brown Rosser's wife was upset. Momma offered to buy the pig, but his wife said, "That wouldn't take the hurt off the pig." Momma said, "Well, what do you want me to do, kill the boy?"

The woman kept the pig and nursed the leg until it got well. She was the darkest-skinned woman I ever saw. Momma was light. That woman didn't like her for that. They were the blackest folks I ever saw, the whole family.

But I know they weren't any kin to us because their son Henry married Aunt Leila's daughter Margaret. It was a case where a Rosser married a Rosser, and they weren't related. Aunt Leila had that girl before she married, so Margaret went in the name of Rosser.

Kroft was a big un'. He was brown, part hound and, I think, part bulldog; but he was a good hunting dog. He could "set" birds, "jump" rabbits, and "tree" squirrels and opossums. He went hunting with Pa all the time.

Two or three White men had gone hunting with Pa one night and saw how good a hunting dog Kroft was. They offered to buy Kroft from Pa, but he wouldn't sell. Pa had read in the Bible something about "Woe be unto the man who perished with the price of a dog in his pocket" and wouldn't sell him, or any other dog.

That night one of the men said he had a female dog in heat. On the way back, Pa let Kroft go home with that man so he could service the female dog. We never saw Kroft anymore after that. We think that ole White man just stole him.

The last Rosser man I knew of was a *Wilkes Rosser*.[17] He lived out past Macedonia. I don't remember anything else about him.

Chapter 8

FARM LIFE

I didn't know any White men around Lone Oak named Rosser. But I knew the White men there that had money and land. There were four: Bill Lee, the richest one, was the man my daddy rented from. The next was my real grandpa, John Barrow. Then there was Burrell Wise, on down to Homer Culpepper. I'm speaking about those who had big cotton plantations and Black folk working on them, picking cotton and working their fields. Those four men were rich because they raised a lot of cotton.

Bill Lee's name was really William Lee. Everybody called him Bill, and they called his son William. Burrell Wise had the biggest store. His daughter Ruth[18] married Bill Lee's son, and they were the richest folk in the store business for miles around. The store still sits at the crossroads in Lone Oak, where four roads take you to the four "villes."

Those men were concerned that their daughters married men who also had money. Bill had two brothers, Albert Lee and Cape Lee. Albert had a daughter who had a secret. I probably ought not to tell it, but I heard grown folk talking about it. When she was a teenager, she had sex with a Colored boy named Luther Addie.[19] He was a handsome young Negro fellow, a few years older than me. He drove the surrey for Albert Lee. After they got a car, he drove the car.

When her uncle, Bill Lee's other brother, Cape Lee found out, he wanted to kill Luther Addie for having an affair with a White woman. But Bill Lee told him, "If you do that, everybody will find out, and she won't be able to marry." The word would get out, and she wouldn't be able to get married in his society. So he didn't kill Luther. That girl did marry a wealthy White boy from West Point or someplace. His family owned the Downey Mill.

Cotton was the only money crop. It was tough back there then. We had to scuffle to get the "man's" cotton out to have a place to live. There was that third and fourth business. You had to get out three bales of cotton to make it. You got the *third* bale. The boss got the *first two*. He had furnished the land, animals, equipment, feed, fertilizer, seeds, and so forth. All you furnished was the work as a sharecropper.

The White landowners had a way of handling slaves when they freed them. It was decided at the hearing of the Emancipation Proclamation that they'd have a way of handling the slaves. They had a system for sharecropping. We didn't have shackles, but we had to give a third of our cotton and a fourth of every other crop. Even to the greens in the garden, peas, potatoes, and corn.

You couldn't even get to keep all of your roasting ears when you grew corn. Corn was an important crop as food for people and feed for animals. The crib was a place in the barn where you stored the corn. Most barns were the same. On the left was a shelter for the buggy. On the right was a shelter for the wagon.

On the back end is a stall for the mules and cows opening to a pen, but they soon found out their stall. And that's where they'd go because that's where you put their feed. The middle of the barn, which had a floor, was the crib.

The way the "fourths" worked was like this: When we would pull the corn, we would fill up our crib the first time. The second time we would fill up our crib. The third time we filled up our crib, and the *fourth* time we had to haul it to the White man and fill up his crib or put it in his barn.

Farming was a hard life. I had jobs to do when I was a little bitty fellow, about four or five years old. I'd have to pick up chips, feed the chickens, bring in wood for the fireplace, and gather eggs. Mae showed me how, and I'd do those things till I got bigger.

When I was about seven or eight, then I'd feed the mules and cows and pigs and chickens. I had something to do all the time. To feed the chickens, you had to make up some dough with meal and water.

There was an old rooster I used to have that would run himself to death and not get anything to eat. I would throw a batch here, he'd run here. I'd throw another batch there, he'd run there. Hahaha!

The slop bucket was always heavy when it was full. Most of the time it was pretty full. That made slopping the hogs a little tough. They would eat anything you put in their trough and their pen. Feeding the pigs and chickens was easy. Every day you had to get corn out of the crib and oats out of the upstairs in the barn to feed the mules, carry cows to pasture, go get the cows, milk the cows, and draw water for mules and cows.

I had so many sisters I didn't have to do much cooking or washing clothes. I'd just draw the water and make the fires. You had to cut stovewood, see that they had enough, stack it up so it could dry, and put some up under the house so you would always have dry wood. It was wood for the stove called sto' wood. Those chores were all through the day.

Drawing water was the biggest job! Of course, you'd have to draw some for the kitchen. That went according to what they needed in the house. But mules and cows drank a *lot* of water *every day*. You could have water drawn for mules, the cows would come along and drink it all up and go to the barn. Then you'd have to draw some more water.

Drawing water meant getting four or five buckets of water to fill up a drinking trough. These were the big wooden buckets that held about two or three gallons of water. To get them all watered, you'd have to draw about ten to twelve buckets.

The trough was beside the well. The well was like a wooden box sitting on the ground with a big opening in it. It had a lid on top of the opening that you had to take off, unwhirl the rope, and let the bucket down. You had to let the bucket down about twenty or thirty or more feet, let the bucket fill up, then haul it up.

At St. Paul, the well was just about eight or ten feet to the water. It was closer to sea level. They called it film water. The well at the Latimer place where we lived was about fifty feet deep because it was so far above sea level. You see, water seeks its own level. When it would rain, the surface water would collect then slowly sink down until it was back at sea level. So if you were up high, the well had to be pretty deep. The farther down the bucket had to go, the heavier it was.

The bucket was tied to a rope on a pulley that you turned by a handle. When it got to the top where you could reach it, you pulled the bucket up by hand and lifted it over the edge of the well to empty it.

That bucket was heavy and hard to crank. Sometime you would draw the bucket almost to the top then let it slip. That crank would unwind so fast it'd knock you down, and that bucket would fall back in the well. It happened to me a couple of times.

What was really terrible was when the bucket would fall back down and the whole thing would go off the pulley and into the water. When all the rope fell in the well, you had to get a steelyard. A steelyard was used for weighing big things. It was a long pole that had something called a pea attached to it. The pea had three hooks on it, two little and one big.

You took a rope, tied it around the steelyard, put a loop around the long handle, and let the pea down. That bucket would fall through a few more feet to the bottom. You couldn't see it and had to fish around down in the well water until you could hook on to the handle of the bucket.

When the bucket caught one of those three hooks, you pulled it up and put its rope back in the whirl. The rope tied to the bucket handle was nailed to the whirl or pulley with a tenpenny nail. The nail was about two and a half inches long. When you got that straight, then you had to finish drawing the water. Oh Lawdy, it was tough back then.

steel·yard (stēl′yärd′, stil′yərd) *n.* [STEEL + YARD[1] (in obs. sense of "rod, bar")] a balance or scale consisting of a metal arm suspended off center from above: the object to be weighed is hung from the shorter end, and a sliding weight is moved along the graduated longer end until the whole arm balances.

Source: *Webster's New World Dictionary of the American Language*, Second College Edition. David B. Guralnik, Editor in Chief. The World Publishing Company, New York, NY. (1970) p.1394

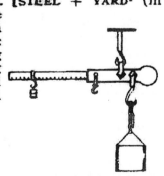

STEELYARD

Steelyard excerpt copied from Webster's New World Dictionary

Many a day it would be so dark we had to light the lantern to see how to draw water and feed the mules. My daddy believed you should work from "can't to can't." That meant it's so dark in the morning you *can't* see, and you work until you *can't* see at night.

By the time I was ten years old, I could plow like a man. So that was my job, along with taking care of the mules. On Sunday I had to put mules in the pasture so they could eat grass. If a cow got in heat, I'd have to take her across the road to a bull and have him service her.

When I got to the sixth grade, I had to drop out of school and plow full-time. Pa would hire me out to other farms, and I'd plow our land after I got through with theirs. I'd eat a bite of supper, maybe corn bread and potlikker, then go to bed. My day would start at four thirty the next morning.

We didn't have an alarm clock, but he would always wake up early in the morning. I never did know how he could do it. It was a hard life. We'd be relieved to hear somebody died. We would be sad for the family, but Pa would go to the funeral. And there'd be a lot to eat afterward.

I can remember what I am saying. When I got up to where I could plow so well, and William could plow so well, the ole man was still able to plow. He hired me out to White farmers for plowing, and they were paying eighty-five cents a day. I was plowing with about nine or ten Black men. Charlie Steward Colly hired them. He had bottomland near where we lived.

When they decided to give the men a dollar a day, the little ole straw foreman, Dave Plant, told me that they were going to give the men a dollar a day and they were going to let that boy plow on for eighty-five cents a day. I quit.

He needed me because a mule was standing idle. He had to come back and get me. He told me, "You have plowed as well as the men. You will get a dollar a day like them," and I went back to plowing for them.

Between the ages of thirteen and seventeen, Maynard Sewell and me plowed with the men getting a dollar a day. Maynard went to Ohio and got four dollars a day, while I was still down here getting the dollar a day.

It was ole Dave Plant's dog that went mad. That dog came up into our yard and chased Mamie into the kitchen and came on in our house after her. She climbed up on the cabinet to get away from him. He had his paws on the window snapping at her. She was screaming. I think it was Amy who was screaming and carrying on too. Pa got his shotgun, and when his shot rang out, that dog fell on out the window. Pa was a good shot; he never missed.

I drove for ole man Marvin Willingham. He owned all the land on that Grantville road from several blocks outside Lone Oak on up to Aunt Mitt Carter's house. People called him "chintzy" because he was so economical he would go to war for a dime before he'd spend it.

I had to run his tractor. He didn't trust anybody to cut his wheat grain but me. As the sap goes out, it gets a small stem. When grains ripen, they are heavy on a little stem.

It took two men to get it done. He had a reaper and a binder and needed someone to drive his tractor right. The tractor cut the wheat and let it fall on an apron. The apron brought the wheat and stacked it up.

When it was full, it had a trap to fasten the wheat together. If the wheat wasn't long enough, the machine would turn it out loose. You had to be alert, so he handled it. He'd stack it in front of him standing until it would make a shaft. He would give me the signal when to stop. Whoever drove the tractor had to be alert to know just when to stop, go right, or go left. Nobody could do that to suit him but me.

The only time I remember working with vetch was for him. I had to plant it; then after it came up and was ready, I had to turn it over into the ground. I didn't have to do anything else to it.

He had four boys. When I went to work for him, the oldest one was off in school and the baby boy, Russell, wasn't big enough to work. I had to supervise the other two. Albert was about the same age as me, and Marvin, Jr., was younger.

One time I told Albert if he didn't do the work, I was going to break his ass. He laughed and said, "It's unbreakable." I said, "I'll knock it out of place, then." We got along. I showed them how to eat currants. There were some bushes on their land, and the berries were ripe. They didn't know you could eat them. Back then folk were scared something might be poisonous and didn't eat a lot of different things. One of the boys tasted one, and his brother asked him, "Is it sweet?" He said, "No, they ain't sweet, but they good as the devil."

The city of Lone Oak gave me a job cutting grass on the streets with a hoe. It was ten cents an hour. I made a dollar in one day, my first day. I had previously been hired out to plow for White farmers. I guess Albert Lee didn't worry about me because I was the only one he wanted to drive his tractor. I was driving his tractor, cutting his wheat and oats.

During these years Pa would give me a patch in which I could raise cotton. I remember trying to get out my bale of cotton. The boll weevils were getting to be a problem. I had to take care of my little cotton patch in my spare time. I had to plow and do all the other work on the farm first. I had to pick cotton, dig potatoes and peanuts, and so forth.

I would keep my little patch clean and make my bale, but all I would get out of it was a ten-cent "harp." I was discouraged. After five years, when he drove around over the plantation and told me where I could have a cotton patch, I told him I didn't want one. He cried because he knew I was planning to leave home.

There was a building boom in Ohio. At that time Negro boys were leaving Georgia and the South. They were going up to northern states to get work. Since the turn of the century, these southerners would go up there building houses and digging ditches. He thought I was fixing to go off to Dayton, Ohio, and work in those ditches and catch my death of cold.

He talked it over with Momma. She agreed that he had never given me as much as a dollar from my cotton patch. Later he approached me and told me what he'd give me if I would stay home and help him make a crop. That is why he willed me that Remington shotgun. That is all he had. It was a new gun. I remember when he bought it.

He bought it in 1912, right after the *Titanic* went down. He paid forty dollars for it, and now it costs about four hundred dollars. There were seven companies that made automatic shotguns, and Browning made an automatic rifle, but Remington was the best.

He told me that if I would stay there and help him make a crop, he would give me that shotgun. He told me I'd have to let him keep it till he died. And when he died, William took it down from over his fireplace and brought it and put it in my car. That is why I have it now.

To make a long story short, I stayed. Pa finally gave me a cotton patch on halves. He told me I would get half of what the cotton brought, even if he did not get out of debt. I made a bale of cotton. He gave me that half of the bale of cotton, which was forty-eight dollars and fifty cents.

I didn't have a bank account. Pa didn't have a bank account. When you sold cotton, they handed you cash money. Working with some White man as farm labor, I made that out to be fifty dollars. That was more money than I had ever had at one time.

Endnotes, Meriwether County, 1906-1923

1. *Bro* was pronounced "bruh," like the first syllable in the word "brother."
2. It may have been a 1908 T-Model Ford.
3. Full names, nicknames, and dates of birth and death for sisters and brothers in birth order are Jennie Mae "Mae" Rosser Jackson Carter, July 6, 1902-May 1, 1975 ; Elmira "Mire" Rosser Blunt Thompson, November 5, 1903-November 1, 1987; and Leila B. Rosser, 1905-1905, were born before JR. The ones born after him were Mamie Frizzell "Frizz" Rosser Johnson Shuffer, December 11, 1908-April 16, 2000; William Henry "Bill" Rosser, February 28, 1911-February 1, 1982; Amy Brown "Top" Rosser Black, January 17, 1913-February 1, 1983; Joe Nathan "Joe" Rosser, January 20, 1915-August 29, 1989; Ellis Holsey Rosser, November 8, 1916-March 31, 1974; John Irvin "JI" Rosser, June 13, 1919-June 9, 1999; and Laura Robert "Bobbie" Rosser Anderson, August 1, 1921-January 14, 2002.
4. Passages from the Bible that were likely misinterpreted and isolated from context are Deuteronomy 14:29 and James 1:27 from the King James Version most preferred during that era.
5. Clifford Ellis, Martha Ellis, and Inez Ellis (who died in infancy), Freeman & Freeman (1985, p14).
6. Eudora Booker's mother was Willie Rosser Booker, another daughter of Anna Hutchin Rosser. Her father was James F. Booker (Ibid, p7).

7. A children's game common to early USA, thought to be a forerunner to baseball. It was also know as Cat, Old Cat, One ol' Cat, and Two ol' Cat (www.wikipedia.org).
8. Distance is approximately fifty yards.
9. *Meeting* was another name for a worship service with singing, praying, and preaching.
10. Spelling of name not verified. Could have been Cuthbret or Culbreth.
11. "I think the Black Rossers may have migrated up from a South Georgia plantation after they were freed from slavery. They all belonged to the same slave-owner, and that's how they got that name. I never came across any White Rossers in Meriwether County" (Personal communication, Minnie Freeman, July 2003).
12. William Rosser and his first wife, Claria Rosser, had one son, John Henry "Peck" Rosser, and two daughters, Symanthia and Fannie. He and his second wife had five children: Dora, Fannie, Emmus, Marvin, and Sun (Ibid, p8).
13. Warner Rosser married Babe Rosser. They had the one daughter and four sons as listed (Ibid, p8a).
14. Walt Rosser, son of Warner Rosser, married Lillie Rosser and had eleven children (Ibid, p8a).
15. Linzie Rosser was husband to Anna Hutchin Rosser, who was sister to Jennie Hutchin Ellis. Their children included Fletcher Blount and five daughters of Anna and Linzie Rosser: Stella Rosser Addie, Willie Rosser Booker, Mollie Rosser Smith, Arvella Rosser Sewell, and Ellie Rosser Shelly. Although JR Rosser, Sr., did not know about all six of these children of Anna, he recognized the names of two of her grandchildren.
16. John Brown Rosser's wife was Emma Freeman, daughter of Aunt Deena and Uncle Joe (Ibid, pp.1F, 24). Likely the fishing hole was named after this Aunt Deena since she was the only woman by that name in the area.
17. "My father was Pomp Ector. My mother was Dora Rosser Ector. She was a Rosser, but they weren't any kin to Pa and Joe. My grandfather was a Rosser from up above Zion Hill, and some of them were in Greenville. He was named Wilkes Rosser" (Personal communication, July 2003, with Bessie Rosser, born December 10, 1917, died 2005, wife of Joe Nathan Rosser, lived her entire life in Grantville).
18. Annie Ruth Wise Lee (July 14, 1899-April 25, 1989) and husband William Perry Lee (November 25, 1896-November 28, 1960), Allen Lee Memorial Cemetery.
19. Linzie and Anna Hutchin Rosser's daughter, Stella Rosser, married Walt Addie. They had eleven children. There were eight daughters and three sons: Luther, Leroy, and Walt, Jr. (Ibid, p6).

FORSYTH AND ATLANTA

Year	Age	Grade	Events	
1924	18	6	SY 1923-1924	School of Agriculture and Mechanic Arts[1]
1925	19	7	SY 1924-1925	Forsyth
1926	20	8	SY 1925-1926	Forsyth
1927	21	9	SY 1926-1927	Forsyth
1928	22	10	SY 1927-1928	Forsyth
1929	23	11	SY 1928-1929	Forsyth
1930	24	11	AY 1929-1930	Morris Brown Academy one-half-year academic year
1931	25	12	AY 1930-1931	High School completion, May First teaching job, June-July Mother's death

Chapter 9

BOARDING SCHOOL

My mother was interested in my getting more education. In Meriwether County there was no school for Blacks beyond grade seven. She wrote to a man at Forsyth who had lived in our county. We had heard about a school for Black boys and girls that had passed the seventh grade. I called it a reform school because that's all I knew about. When boys went away, it was to reform school. It was almost like breaking the law for a "nigger" boy to be in the community with no job.

In September 1923, at age seventeen, I was off to boarding school at Forsyth. My momma prolonged my life. In fact, she was the one who first took me down there and enrolled me. It costs ten cents for each of us to go one way.

We rode three trains. From Grantville to Newnan we rode the *West Point Line* that went to Atlanta. From Newnan to Griffin we got on a train coming from Chattanooga to Griffin, which went back to Chattanooga. From Griffin to Forsyth we rode the *Nancy Hanks*. That was part of the Great Southern Line that went from Atlanta to Florida.

It was a state institution. A principal named William Merida Hubbard ran it. It was just like going off to college is now. For somebody who hadn't been used to indoor toilets or electricity, it was quite an experience. It was the first time I remember sleeping on springs. We weren't quarantined on the farm campus. It was just a gathering place where they sent boys who got into trouble, only we were not in trouble.

There were four institutions of this type that made history. The four *H*s ran them: Hubert, Hunt, Hubbard, and Holly. They established the schools. There was one at Savannah later known as Savannah State College. Hubert was at Savannah. One at Fort Valley became Fort Valley State College. Hunt was there. One at Albany became Albany State College. That's where Holly was. And one was at Forsyth where I went, which later closed. These schools attracted many of the Black boys and girls that wanted to go above the seventh grade. They were from South Georgia, Alabama, Louisiana, and so forth.

When I first went to Forsyth, I paid the tuition of nine dollars for the first month. I was on a work program and was such a valuable person; that was the only tuition I paid. I had been driving tractors four or five years. It was an agricultural school. Because of the fact that I knew how to plow and do farmwork, including driving tractors, the principal told me that if I would do that for the next seven or eight years, he would let me go to school tuition free. That was the situation of my being at that school. I took up the work. I worked so well that I drove the school truck.

The first year I didn't come home until the summer. Ole man Willingham's wheat was about to fall. He'd let it stay till I got there. I don't know what happened to his wheat after that year. He was old. He may have given up planting it. That was the only work I did for anybody else in Meriwether County. My daddy was farming, and there was a lot to be done. I helped out all I could until I went back to school. I was the first one from Lone Oak community to go to the Forsyth School. After I went, several others came along.

Money was so scarce until there weren't any allowances for any children in my daddy's house. I couldn't have money in my pocket like other boys had. I couldn't buy a bicycle like the other boys bought. I was a grown man when I decided not to ever be caught without money in my pocket.

I guess indirectly, in spite of the situation, I admired Uncle John Henry because he had money. He had some tough ideas about work. He told me this story about a man that was working for him. He told him, "I'm running this 'cat' and you're hired to hold his tail. If he farts and you fall over, I'll raise your wages." Hahahahaha!

Cat meant this is my job. The moral of the story is I'm doing my own business, and if it's too hard for you, let me know, but don't come telling me how to do my job. You see, Uncle John Henry had been cooking syrup for years. That's what he was doing when he told the fellow this, cooking syrup.

It was strange. When I had to go off to school and could get an education, I guess my uncle didn't really approve of me going to school. When I left, there was a big demand for "nigger boys" to do farmwork. Well, I had hired out to some White men, driving their tractors. It was a farming community, and I could handle it well. He felt like I ought to have stayed there and worked driving those men's tractors instead of going to school. That is the truth about it. I found out because he told me himself. He said, "Son, maybe you ought to stop school and work." I didn't say anything to him. I worked, but I didn't stop school.

I had to do elementary studies in the sixth and seventh grades. Practically the same classes were offered day and night. I went to day school and night school. If there was work to do at daytime, I had to do that and go to classes at night.

When it rained, I went to school every *day* I could. I went to class every evening for night school. The teacher in the day was Nellie Thrash. Professor Samuel Blanton Pride was a teacher for the day and night school. I received two certificates of graduation from grade seven, one from the day school and one from the night school.

Professor Pride was my inspiration. He was the dean of the boys boarding department. I liked him. I named my second-born son after him. Pride looked like a White man with pretty hair. He dressed nicely. He was a good teacher. He taught me Latin and algebra. He was nice to me. He taught me baseball, taught me how to pitch, and he was our basketball coach too. He was nice to us out on the ball field. I sort of aped him when I was teaching school. He had graduated from Shaw University.

Teachers at Forsyth were graduates of Savannah State, Shaw University, and several other places. They came there because it was a teaching job, and they had taken courses they could teach. They didn't meet any special requirements. They didn't have to have a college degree. In fact, I think some of them taught who did *not* have a college degree.

In high school, I studied Latin. It was a required high school subject. Algebra and geometry were also required subjects, along with English, geography, history, and science. When I learned to say something that most other people on the street could not say, I appreciated it. When in algebra, I learned to let x equal the unknown and solve the problem to find the number; I was proud of learning those things. The other boys in my community didn't know that.

The music teacher, Miss Aquilla Jones,[2] was scrambling over the students to try to get her course together. I could sing, so I joined in with them. I sang in the commencement programs. When Audrey did her practice teaching in music at Ballard-Hudson Senior High School in Macon, her supervising teacher was this same Aquilla Jones. She recognized the name and remembered me.

We had a band at Forsyth. We got some instruments from Georgia Tech. When the state gave them new instruments, they gave us their old ones. We had two cornets—some called them trumpets. We had a french horn, a trombone, a bass horn, and some drums. The bandleader's name was Edward L. Farley. No one could play the french horn but him. He assigned the other instruments. Frank Tucker had a cornet, Frank Springer had a cornet, and I had the bass horn. A fellow named Hartfield had trombone, John Hudson had the snare drum, and Charlie Mattox[3] had the bass drum.

We got those instruments about the first of December. I had never seen a horn and didn't know how to play one. It took days to learn how to make a sound. After a week, I could make sounds come out. I couldn't read music, but I could read what he put on the blackboard. First, he put some figures for notes and things on the blackboard. And when we got to playing it, it was "Silent Night." That was the first song that we learned how to play. It's still my favorite Christmas carol.

The band played for some programs. I shook hands with R. J. Moton.[4] He was sort of drumming up the NAACP because he was sort of in a fight with Alabama. The high school Richard graduated from, in Birmingham, was named after him. Anyhow, he came to our school.

It was a big deal. He was a guest speaker. We didn't have an auditorium. We had had a burnout. The city of Forsyth let us meet at the courthouse. Our band marched from school to the courthouse. Some of us had never been there.

Farley was a graduate of Talledega College in Alabama. It's south of Anniston. He was also my French instructor. He asked me what I was planning to be. I wanted to be a doctor. He told me I would have to finish high school, then finish college, then finish medical school, and spend years as an intern and a resident. He told me I would be too old by the time I could complete the training to be a medical doctor.

I thought I could cure my momma, but I gave up on it after he told me that. Got-dang it! That time came and went, and I still had enough good years left in me. He gave me some bad advice. I tried to get in touch with him when I was about seventy years old. I wrote to Talledega College and got the address. He lived in Detroit. I wrote him, but the letter came back.

Momma was influential in my staying in school. She was my all in all. I remember quite well, every time she would write me, she would close out with "Don't forget to pray." Once she sold some butter and eggs for fifty cents. She sent me a dime of that money. She sewed it up in the corner of a handkerchief and put it in an envelope.

When I got it, I was so proud. I wanted to save it for a rainy day, so to speak. I put it in my trunk at the foot of my bed. When I went to get it, it was gone. I didn't have a lock for my trunk. I felt bad, but I never did tell her somebody stole that dime.

We didn't play football at Forsyth. There were one or two fellows who tried to practice football, but they were never able to get a team to play anybody else. I remember once they were running the ball. The fellow had the ball; someone tackled him, and he fell on the ball and it burst. It sounded like somebody had shot him. You know the balls had bladders in them. I understood they later got a real football program started.

When I first played basketball, I was a guard, but so many times I was the high-scoring man. I was shooting long shots then. I was shooting "three-pointers" long before they gave a count of three points. Guards weren't supposed to go too much into the forward area of the court. I would get close enough and shoot. I was not the only one who made long shots.

I remember when the Hawks[5] had a fellow who could shoot long shots before they recognized them as counting three points. Most of my shots were two-hand set shots, but I had a one-hand jump shot. They were discouraged by the coach because the ball could be so easily stolen.

We played other schools, most of which were in Macon, just twenty-five miles south of Forsyth. There were several high schools in Macon: We played Ballard, Hudson, and Central City. They had some pretty good teams. Central City was a church-supported school, not really a seminary. Boys who wanted to preach went there. We had trouble with them because there were many guys on the team who were older than we were.

We played Canton, Temple, and one other little town out from Douglasville. We didn't play Douglasville because the school there was bigger and in another class. All of the teachers at Forsyth had Ford touring cars that were kind of like convertibles. Those cars are how we got to the games. There were no seat belts. You could pile 'em in long as you could get 'em in. The law had nothing to do with you unless you had a wreck.

One of the other players drove the professor's car until I finally got to driving. Sid Lambert had a Ford coupe that could carry all six of the girls in one car. One of them was grumbling about somebody sitting on a lap. I said, "If you don't want to go, get out." She said, "Yes, Professor Rosser, I want to go."

My brother-in-law Allen Jackson had a 1925 Ford touring car. It was the first automobile I drove. He had stopped it in the middle of the road to talk. I was driving a wagon. He was going to move it, but I stopped him. I got out and started it and drove it out of my way, then got back in the wagon and went on.

I drove two of the professors' cars. One, named Deadwiler, was a farm agent. He had me to drive his car when he had to send for people. He would go home on weekends, and I would drive him to Athens. Once he said, "When you get down to the fork in the road, turn right." I started to turn left. He said, "Whoa, wait a minute. You'd better get out here and throw a rock. Do you know what right is?" Well, I did know. That was where I experienced learning about paying attention.

John Deadwiler became a good friend of mine. Once we had only syrup and corn bread for three months at school. After students were so weak they were falling down, they went on "strike" and threw it at people who came into the dining hall. Deadwiler drove us to town, got biscuits and rolls from the bakery and sardines in tomato sauce. We ate our fill!

I was driving Professor Collins's car when I was stopped by the police. I was going through a little town of Woodbury, in Meriwether County. They had a street labeled Through Street. I was going down it in the rain. A White man was coming toward me from a side street, and it looked like he wasn't going to stop. I speeded up and went by him so he could get by and to keep him from hitting me.

He made a case against me. He followed me out to the next filling station and told them that I like to have run over him. He was driving a Ford, as I was. After that, they had a warrant out for me. Professor Collins carried me back over there for the hearing. I drove the car, but he directed me.

That was my first encounter with the law. The hearing was before the mayor and town council. They didn't have a jury for some little "nigger" boy. I explained to them that I didn't stop for that man to come on because I was on the through street. He looked like he was going to come into the "through street" from the side street that intersected. If I had put on my brakes, I may have slid into him and caused a wreck.

The real cause for the situation was that he saw I was Black and thought I ought to stop, but I speeded up to get by him without hitting him. I didn't know then, like I know now, that everyone in the court could see that the White man was at fault. They dismissed the charge against me.

Professor Collins was a lawyer. He thanked the court for dismissing the case. He said he was a Meriwether Countian, and he was a good boy. And like all Meriwether Countians would not break the law. He took a fellow's typewriter and kept it for ten years. Then took him to court and beat him out of it. Sometime being right pays off to your advantage, sometime it doesn't.

I never had time to chase women. I never was interested in doing that and only remember two that were interested in trying to "hook" me. One summer I stayed at the school. It was only about fifteen of us. For some reason I had to do the cooking and a lot of farmwork.

There was a nice-looking, sort of matured, brown-skinned woman who was in Forsyth visiting some acquaintances. She was going back to Florida and wanted me to go with her. She was married and said she would put me up somewhere else so we could see each other.

She bought me a ticket, and I agreed to go. I had so much work to do I lost track of time. When I remembered it, I heard the train whistle blowing, and there was no way I could catch that train. That's how I got out of it. It would have been terrible to get mixed up with her.

I had another close call the last couple of years I was in Forsyth. A little girl named Mabel Brown was a year ahead of me in school. She lived on the northeast side of town and we lived on the northwest side. I had an engagement with her, and I went over there across town and talked with her. It was about two miles I had to walk. Her mother was dead. She lived with her grandmother. When the woman said it was nine o'clock, I would have to leave.

After about three years, we were courting hot and heavy. I was to go from Atlanta to Marietta to meet her family. It was some sort of dinner. I always had to work, and when I got on the train, I went to sleep.

When I woke up, I was in Dallas. By the time I got a ticket and got back to Marietta, it was late. She wanted me to meet her uncles and aunts and they were all gone. Shortly after that, we broke up.

I was kind of lucky. Had I gotten off the train in time, I probably would have married her. Mabel got interested in a preacher with a car and married him. When I was at Morris Brown, I thought she came to see me, but she came to see him.

After three or four years, my sister Mamie was at the school. She worked in the kitchen. She was to work for half of her tuition. Pap died while Mamie and I were both in boarding school. Reverend South didn't let us borrow his car to come home to the funeral because Mamie and Mrs. South didn't get along.

The reverend liked me because I escorted young ladies to the Kynett ME Church,[6] where he was pastor. They would put a few pennies in the collection plate, or a dime or so. And back then, every dime counted. Professor Stough let us borrow his car to come home to the funeral.

Old man Hubbard had told me, "You drive the truck around and do the hauling we have to do, and you can go to school." The other boys and I would go in to get the little pay, ten cents an hour for the number of hours we worked over forty hours a week. After I did it for two or three years, we made an agreement. I told him to hold mine so that in my senior year I would not have to work and go to school.

He agreed to that, but he didn't keep his word. Hubbard called me in and told me that he felt like the school was doing too much for one family by letting two out of the same family work their way through school. I said, "Oh, I see your trouble now. Don't bother Mamie, I will look out for myself." When he knew anything, I had gone to Atlanta.

Fortunately, Mrs. Rembert,[7] who was the matron at Forsyth, had been a matron at Morris Brown. She contacted the president of Morris Brown and told him about my situation. When she wrote President Fountain and told him about me, he wrote her back, "Send the young man on, I'll take care of him." That is how I got to Morris Brown on a scholarship.

Mrs. Rembert also asked teachers to donate money to help me. They donated eighteen dollars to pay for the train fare to Atlanta. That is all of the money that I have had given to me. I caught a train and went to Atlanta.

After I left Forsyth, it grew to become a junior college. My sister, Mamie, stayed there. In addition to finishing the high school, she was one of the first graduates from the junior college.

Chapter 10

HIGH SCHOOL

I went to Morris Brown in the middle of my junior year. I was there a year and a half. The curriculum was different from that at Forsyth. Morris Brown's high school department was connected to a college department. They called it an academy and called the program "academic." They had a seminary, and fortunately, they had boarding students.

The college had devotions where an instrumental quartet played. I went to those devotions. I was sitting up there, a little high school student. It was a long time before they found out that I was a "prep" student.

Because I went in the middle of the school term, I had to take the preacher course. I had to take two seminary courses. We had to study the Bible and memorize certain parts. I remember some passages of scripture that we had to memorize, such as David's charge to Solomon:[8]

> Now the days of David drew nigh that he should die; and he charged Solomon his son, saying, I go the way of all the earth. Be thou strong therefore, and show thyself a man; and, keep the charge of the Lord thy God, to walk in his ways, to keep his statutes, and his commandments, and his judgments, and his testimonies, as it is written in the law of Moses, that thou mayest prosper in all that thou doest, and whithersoever thou turneth thyself.[9]

Also, I remember when Israel asked for a king:[10]

> Then all the elders of Israel gathered themselves together and came to Samuel unto Ramah, and said unto him, Behold, thou art old, and thy sons walk not in thy ways: now make us a king to judge us like all the nations.

The passages of scripture were important facets of my life. They helped me in my thinking more than in my advising.

The scriptures were very helpful. In fact, I could use David's charge to Solomon in my business: When all the elders of Israel were gathered and came unto Samuel at Ramah, and told him to "make us a king, just like all the nations around here." I used that quite frequently to show people how some folk can get too high for their positions and feel like they need something else other than what God had provided.

In this scripture Samuel is a judge. His sons stand to inherit his judgeship because they were under the system of the "birthright," but they are corrupt, and it seems they won't serve the people well. If judges could be satisfied with what God had for them, they would be better judges and render good judgments that would satisfy God's plan. Then, people would respect them and consider their advice. A lesson here for people who have high positions is to act *right*. When people don't follow God's teaching, they delay his plan.

As I got older, I taught Sunday school all along. One of the first times I taught, the subject was "Who Is Eligible to Pray." I told them Daniel's prayer was so strong that it locked the lion's mouth. Moses's prayer was so strong it parted the Red Sea. And I went on with that.

All I remember I was ever taught in Sunday school back in Lone Oak was to respect folk, grown folk, anyway. My Sunday school teacher was named Dee Sewell. His class was mostly boys. They called him Sugar in the community. After I went to school in Forsyth, his son went there and got into it with some of those Forsyth girls. He married Mary Brown.

While attending Morris Brown, I delivered newspapers in Atlanta. My classmate, a buddy of mine who had been there all the high school years of that class, had a twenty-five dollar bond with the *Atlanta Constitution*. They sent him fifty-two papers out there. He kept the bond, but he went to Piedmont Hospital to work, and he got me to carry his papers to subscribers. For several weeks he let me throw papers, and when he collected money for them, he'd pay me.

The bunch of papers was dropped off at Edgewood Avenue by streetcar, beside the Constitution building. I picked them up and delivered them to the customers on Edgewood Avenue and around up there. It was around the area where the guy was rescued by helicopter recently.[11]

The Fulton Bag Company burned up, and a fellow was up on the scaffold, and there was no way for him to get out. And the fire department fellow flew over in a helicopter. The Fulton Bag Company was in the area where I delivered papers—that's the reason I mentioned it.

I was firing the boiler at a men's dormitory, called Flipper's Temple. I don't think it still exists because Morris Brown moved to the west side years ago. That was where I got burned. The fire was choked up; it wouldn't burn much. I stirred around in it and got the fire to catch up.

It blasted up and blasted fire gates to the flue. It blew open and blew up at me. It caught my face and head afire. I attribute being bald headed early to this incident. My daddy was not bald when he died at age eighty-five.

I went through some rough times. Once I was out of money on Wednesday and didn't get any more money until Sunday. I would buy lots of stale donuts. You could get twelve for a nickel. They would be dried out. But a lot of us ate them.

I had my tonsils removed at Grady Hospital in Atlanta when I was twenty-four years old. That's the only reason I am still living. I had to stay in the hospital over a week. I had bad tonsils, and they were so big.

A couple of times I sort of brought things on myself by not thinking about the consequences of my actions. One time when I was delivering laundry, a man paid me six dollars and the wind took it away. I wished I had put it in my pocket, but I had practiced not to put any of the laundry money in my pocket. Oh boy!

One other time I stayed up all night playing checkers until the sun came up, and I was supposed to have read. I wished I had read that passage because checkers weren't worth *not knowing* the answer to questions the next day. It's hard to catch up when you get behind.

I got this ring with a green stone in Atlanta too. A fellow came up to me on the street. He showed it to me and was telling me it was a diamond and would cut glass. He was cleaning up at Rich's department store and said he found it. I believe he stole it. He wanted five dollars for it. I gave him two dollars. He said I could give him the other three dollars if I ever ran into him again. Of course, I never saw him after that.

I tried not to be overly influenced by my peers. I met up with a preacher called Battleax Davis. Instead of living in the dormitory, we decided we would live on the street, but that wasn't doing it. I got hungry. So I had to go back and pay the room and board for all except that two weeks.

I met another preacher named Shal-zeek from South Africa. He came over here more or less as a missionary. He was in the AME[12] Church. I learned from him that a whole lot of menfolk from his group thought they were priests of some sort. And they had deep dominion over the women. Women didn't have *any*, or many, rights. I always thought it was a shame that women couldn't vote before they could vote by law.

I got an athletic scholarship to Morris Brown. That was based on my baseball ability, and I got a varsity uniform to start. I was an ace pitcher on the baseball team. They called me "steel arm." I was known as a "fast-ball pitcher" because I could throw an overhand ball so fast. Due to the fact that I had been out of school for five years, I was older than most of my classmates, and I was a little stronger.

I struck out twenty-one men in one game. Ours was a school team, but the city we were playing had a team of old men. There were just six men that they had to get out. I didn't have any better sense than to strike them out because I could. The account of my striking out twenty-one batters was in a newspaper article in one of the Macon papers. I kept the little old printed page from the *Macon News* until it turned yellow, and I don't know what happened to it.

You know, I haven't seen that done since. I think I've seen where someone struck out twenty-one men. I think one was the Brooklyn Dodgers by Sandy Kofax. I can't remember the name of the other left-hand pitcher who struck out twenty-one or twenty-two men.

Once Morris Brown went to play the high school at Forsyth. I was being disciplined by my coach, so I didn't get a chance to go back and play my old school. I had slipped out on Sunday and pitched for someone.

I played all of the positions on the baseball team but one, catcher. I couldn't catch because when the fellow would strike at the ball, I would shut my eyes. If they didn't have someone who could catch, there would have been no baseball.

Other than baseball, we just had basketball, football, and tennis. We took part in those because we enjoyed them. We didn't get any credit for them. I was on the little ole football team for a season. We went to Chattanooga to play Mars College. What I remember about it was that's when I saw my first Black policeman.

I had not had any special conflict with police; but I thought that a policeman was someone to be scared of. I saw the Black officer wearing his official police cap. I then began to look at police differently. I later had to learn that the policeman was my friend. I have never been arrested, even to this day.

I went out for basketball, but we had a coach who had to have a "fay" team. All players were half-White or very light, yellow, they were not white. He didn't let anyone who was as black as I was play on his team, except in practice. There were times in practice when we "black ones" would beat his varsity team to death. Fred W. Gunn was his name.

There were several of us who were "too dark" for him. We sat on the bench but did not play in varsity games. Even when he was losing the game, he wouldn't try putting in any of the dark players. There was no one to appeal to. We didn't know anything about protesting.

I don't know if you could call it segregation, but that is what it amounted to. Some Black folk thought just like most White folk. If you were a dark-skinned Black, you "stank," would steal and lie. That's the way the White people think of us, even now.

At that time, I didn't know that I could run for president or sheriff or anything like that. You see, that was about a generation or so after "freedom." Freedom came along about 1865. From 1865 to 1906, when I was born, all of slavery hadn't worn off. To tell the truth about it, it hasn't worn off now. But that was the situation. When I began to learn a little Latin and algebra, I appreciated it so much. This was something the little "nigger boys" on the street didn't know. I felt appreciative.

For a long time, you didn't call anybody "nigger" or "black" if you were light skinned. If a fellow called you black, who was a little lighter than you, you had a fight on your hands. That existed a long time. That was widely known to be the situation. I called it "unwritten rules."[13] For a long time they had "unwritten rules" after they had to integrate, that is, desegregate, schools. They had a lot of policies that were not written but carried out, like a Black couldn't play quarterback.

I remember a story of two Black boys and the railroad. They saw a freight train going up there, straining and spreading all the steam and power. One little boy said, "I wish I was a White boy." The other said, "Why?" The boy said, "So I could drive that train. Look at him just looking out the window." The other Black boy said, "I don't wish I was White, I just wish I could get a chance to run that engine like that ole White man is doing. I'd drive that train, *Black* as I am!"

When I was a senior, my high school had a little ole dance I went to, called a prom. We had a little "orchestra." Saxophone, clarinet and cornets, and a jukebox. I danced with a little girl named Minnie Lee Harvey. We didn't have any tuxes. We wore just what we would wear to church, a shirt and a tie and britches and a coat. I graduated from Morris Brown Academy in May 1931. I finished the high school academic courses and the courses in religion required of all students. I got my diploma. I still have my graduation program with the names of everybody in my class. It was thirteen of us, but only twelve graduated.

My class ring has JTF in it because John Thomas Fuller was the class president. The company responsible for our graduation orders gave the president his ring, and I couldn't afford to buy one. Fuller didn't graduate with our class. It had the year engraved on it, so he gave it to me.

I wanted to stay there in school. I wanted to go on, to take college courses. I left because my clothes were wearing out. I had a checkered pair of pants that I wore to commencement. It was my best pair. And I had a nice jacket. I had to go make a little money, to get me some britches, so I went on to teaching.

Diploma received from Morris Brown upon completing High School

Chapter 11

FIRST TEACHING JOB

When I finally finished Morris Brown Academy, my momma was proud of me. I came home that summer to visit before I went to the job at the school in Mount Lowell. She told me to be nice to the little children. Jenny Mae and Elmira giggled and said, "He'll be nice to the pretty ones." She told me to "be nice to the ugly ones too."

That is the last advice she gave me, and I have tried the rest of my life to do what she told me. She was sick. She had been sick a long time. She had too many children too fast and never healed from some kind of tearing in childbirth.

Mount Lowell was in Carroll County, but it may have been nearer to Waco. There I had to teach in a Baptist church, with a potbelly stove in the middle of the room. It had benches lined up. The little school had a term for two months in the summer.

They let the children come to school between May when cotton was planted, and August, when it needed hoeing. I was the principal of the one-teacher school. *One-teacher* meant I was the only teacher for grades one through seven. I had about twenty-eight students.

There was a pretty little girl in the first grade named Irma Lee. She would say, "I ain' got no sense. You can look at me and tell I ain't got a bit o' sense." I used to tell her she was going to be a beauty queen when she grew up. She'd say, "I ain't gon' be worth nothin'." I'd tell her, "Yes, you are. You may grow up to be Miss Georgia one day."

I knew she was just saying what her stepmother, or maybe grandmother, told her so many times. I heard the woman say it. I felt the same way about her that she was talking about the little girl. She was an ole tan woman, too tan to be light skinned, who hated light-skinned folk.

Irma Lee was friends with Mozelle Graham. They were both in the first grade. Mozelle was as cute as a picture too. She looked like a little doll. I boarded with the Grahams. Mrs. Graham was real dark, but she was a nice-looking lady. I don't remember her first name, but her husband's name was Zenus.

After I had been at Lowell one week, I got a letter from Jennie Mae saying Momma was deathly ill and for me to come home if I wanted to see her alive. The only way I had to get home was in an old one-seat truck with two White men, and I had to sit between them. They passed a jug of rotgut whiskey back and forth and insisted that

I take a swig. I didn't drink whiskey, and I would hold it in my mouth till it stopped burning before I could swallow it.

When I got home, Momma was so weak she couldn't talk, but I think she recognized me because she smiled. She tried to say something. I had to bend over close to her, but I couldn't make out what she was saying. That was the last time I saw her alive. She died later that summer.

It has always bothered me that Momma may have smelled whiskey on my breath and thought I was a drunkard. She would have known it was whiskey. My momma liked whiskey. She could drink it like most folk drink water. When she was a little girl, her mother's husband owned a still. He used to send her down there to get a bucket of the brew, and she would always taste it.

Mrs. Tammy Thomas was a Home Demonstration agent. She would find teachers for the Negro schools in Carroll County, and Haralson too. She got me the job at Lowell. When she carried somebody else there for the next term, the superintendent asked, "What happened to the boy?" Mrs. Thomas told him I went to Bremen where they had a schoolhouse instead of school in a church, and a nine-month school term that was a better-paying job. He said, "I liked him. If I'd a known that, I would have given him twenty-six dollars." My monthly salary at that first job was twenty-five dollars.

I accepted a position in Breman for the fall. It was a raise in pay to thirty dollars a month. Tammy Thomas's husband was Dr. Samuel D. Thomas. They lived in Carrollton, but I met him in Lone Oak. He was my momma's doctor and was the only Black medical doctor for miles around.

I went to Lone Oak[14] to see my sister-in-law Bessie. Her nephew and his wife came up right after I did. The wife was Ivy Dean's daughter Julia Lee Dean Grissom, who called me cousin John Robert. Her mother was Bessie's sister Annie Mae Ector Dean. The nephew was a former student, one that I taught that summer at Lowell.

His name was Marion Grissom. He's the father of Marquis Grissom that played baseball for the Atlanta Braves and who was with the Los Angeles Dodgers last year. Now Marion is seventy years old and a grandfather. He was in second grade at that time. He remembered spelling and counting as the things I taught him. Those were all I could teach the children during that term. He said he always remembered me because I was the first *man* teacher he had ever had.

Chapter 12

AMERICAN UNION RELIEF SOCIETY

Uncle John Henry and nine other men got together and formed a society and named it American Union Relief Society. The AURS was a fraternal society established to bury Black folk. In 1916 they were having a tough time burying the dead. This society helped out on that. It started at Lone Oak, at St. Paul CME Church. Cousin Beodessa Rosser had to help them with the grammar and all. Somewhere in the literature she wrote an ode to the pine tree. In the beginning, the men who started the lodge shook hands around the pine tree.

I remember some of those men that started it. The first worthy grand president was Philly Wright. He served three years. The first worthy grand secretary was Ben J. Rosser. He was from Primrose. Then there were Charles H. Lee, C. W. Upshaw, O. D. Ramsey, and S. D. Wood. Uncle John Henry Ellis was from Lone Oak. He became the second worthy grand president and served until his death.

AURS was a volunteer organization where officers served without pay. There was a deputy secretary for each division. There were four divisions: Division number 1 was at Lone Oak. One of Uncle John Henry's sons was later their deputy. Division number 2 was at Bremen. Division number 3 was at Tallapoosa, and Division number 4 was at Goodwater, Alabama, near Talladega. That deputy lived in Roanoke, Alabama.

All of the members in a division, the division leader, called a deputy, and the grand president met at what we called a division meeting. We had them about four times a year. They would get together and send out calls. I don't know who was initially the head when the divisions were established, due to the rapid expansion of the organization. We had as many as eight thousand to ten thousand members who belonged to it.

We paid our deputy's expenses to the division meetings. There was a tax of three cents per member to cover these costs. This money was collected at the grand lodge meetings, and it paid for the committee work of the grand lodge. That was when two cents would send a letter all the way to Germany.

Dues were based on ten cents. When a member died, surviving members in their division would pay ten cents. The lodge grew so that the ten cents from each member made about three hundred dollars. The family of the person who died would receive three hundred dollars

This was more than enough to bury a person, at that time, pre-World War II. Why, fifty dollars would give a decent burial. Most Blacks were not well-off. Some of these families had never had three hundred dollars and really didn't know what to do with it.

A touchy question we had back then was whether we'd let a White undertaker have a body. That was a no-no. I remember one time a White undertaker questioned the lodge about the big amount, so we had to call it a donation. That was too wide a margin between them and us. There weren't any Black undertakers in the community, and White ones wouldn't fool with it until they learned about the money. When a lodge member died, they would rally to get that body. They knew they were going to get three hundred dollars instead of fifty dollars.

I joined the AURS and went to work with Charlie Byrd and the Tallapoosa division the year I married. Charlie Byrd was president of Lodge number 110 at Tallapoosa. At that time I had no idea that I would become as involved as I later became.

There was a story in the news about an incident in that area of Alabama that was the heart of the AURS. Students were going to have a prom. The father of one of the girls was Black and her mother was White. There was a great commotion. The superintendent cancelled the prom, and they had a whole "to-do" about it because of the race mixing. Since the heart of the AURS operated in that section of the country, we knew a lot about that situation.

Uncle John Henry approved of me working with the AURS. When our burial society met at a grand lodge in Grantville, he asked me to speak. At that time I had graduated from Morris Brown Academy, was teaching school, and had a beautiful wife and two or three beautiful children.

In my speech I demonstrated "the significance of simplicity." I used a straight pin, a simple machine that is straight, smooth and sharp, as a symbol. Once it has been bent, no one can straighten it out. I pointed out that the pin had a point for everything for which it was used. And it has a head.

The purpose of the head is to keep the pin from going too far, the importance of thinking before acting. I concluded by asking the audience, "Wouldn't it be nice if all of us were to let our heads keep us from going too far?"

Mrs. Colly[15] started applauding and got the others to applauding. Charlie Stewart Colly's wife, Mary Colly, an outstanding lady in the women's department, was an invited guest. When she was invited to make remarks, she mentioned that she knew about me as a boy, and that I plowed for her husband. Her statement was "He went off to school and got his education and is now teaching." She asked me for permission to use the pin analogy in her women's program that she was going to conduct in Newnan the next week and in Atlanta. That was an unexpected boost.

Later when our grand secretary died, they put in my name for grand secretary. Uncle John Henry held that up and put Garrett Sellers in there. He got the grand lodge to leave it up to him to appoint a grand secretary. He appointed Garrett Sellers; I guess Sellers discussed it with Uncle John Henry. I don't know that he was in favor of me, but he told Uncle John Henry that I was the best man for the job. So I became worthy grand secretary.

Sellers was a Black undertaker in Atlanta. We had some lodges up that way. In that time, we had several Black funeral homes that were involved with the AURS. I remember one was the Roscoe Jenkins Funeral Home in Newnan. It was owned and operated by Roscoe and Octavia Jenkins.

I was grand secretary, and Uncle John Henry was grand president. We worked that way for thirty years; then he had a stroke and couldn't talk. I had to take that on and prepare a grand lodge program for a year, get out the paperwork and all. I had to learn a whole lot by signals. He'd point out items in the constitution. He'd just stretch and point. I wasn't a mind reader or that kind of stuff, but I had to do the best I could.

I had a guide that I had to follow. I had never made a password until then. You may know, the secret of a password was a number for a letter, like where you had *E*, you'd have a five. I had to spell out a password, put it in code, and send it to the local lodges. It changed every grand lodge. By the next grand lodge you had a new password. I remember the first one ever given to me. It was "work and worship." That would have been in the 1940s or 1950s.

I didn't have any practice, but the lodge members seemed satisfied. After Uncle John Henry died, they elected me as the third worthy grand president. I had to send out the calls. We would have a division meeting and gather up all of the money and pay out to folk. I had to do that for twelve years.

After so long a time, it got to where it wasn't paying enough to bury them. Of course, they had other insurance and things like that. I stayed there until it was too far to go way out in Alabama to collect such a little dues. After they lost so many members, there weren't enough members to pay enough money to go to a division meeting to collect. Therefore, about 1970 the problem was so acute that it wasn't worth fooling with.

Before I became grand president, we had a little scholarship program. We would give a scholarship to a member's son or daughter who finished high school and who was a member of the organization. We gave scholarships to several girls during a period of ten years. When I got ready to retire from it, I recommended a female for president. I felt we ought to receive some benefit from our money since we had been giving scholarships to women all along and had never given a male a scholarship.

I offered Uncle John Henry's daughter, Minnie Ellis Freeman. To start with, I didn't have cooperation among the members of the society, but they finally came around. After so long a time, on my recommendation, they elected her the fourth worthy grand president. That is the way it stands now.

The lodge has gone down since then. When it came to where it wasn't furnishing enough money to be considered the *total* needed for a funeral, because funerals were so high, I was ready to let it alone. They still give that same three hundred dollars when a member dies. Minnie is still trying to carry it on.[16]

They sort of discourage members from staying in because when you have been in it for fifty years, they send you a hundred dollars. They sent mine sometime ago. I thought it had folded then.

Recently I got a call saying I had been transferred to the Bremen Division[17] because the Tallapoosa Division closed. Some woman was trying to carry on for her mother who was the deputy. I was asked to send her a check for thirty-five dollars to cover the dues for one year. That's what she figured it would be to cover the number of deaths expected for the year. I said that it was a waste of money. But I sent her the check.

At a meeting in January 2001, attended by two of the "old" members like myself, they were trying to rebirth the organization. I'm ninety-four years old. I can't do what they would want me to do, like go to meetings and organize folk. I can't do that anymore.

The trouble is that young folk today don't care anything about such things. Since integration came around, young people can take part in any club or organization they choose. An organization like the AURS is not popular anymore.

Many other burial organizations in that community would spring up and go down. They had what was called a UBA Society, *Universal Burial Association*. I don't know what became of it. Several other names, I could possibly recall, where two or three fellows got together and decided they wanted to do something, but when they really found out what it would cost, they dropped it.

A Sick Committee would go around and ask about other folk. They would keep up with it. Someone would be at the home every day when there was a real sick person there. I guess they were more expensive than they were worth because they would eat meals. But at least they were there and would report on the sick.

Before the AURS came about, Black folk had a tough time trying to bury their dead. The Sick Committee would go around and ask folk to give what they could afford. They were scared not to contribute. They knew that their time was coming. And it was awful. You'd have to try to live up to your pledge. I think some folk would pledge just to get rid of it. They just promised it. They never would pay it.

Families had to do the best they could. The bodies were brought to their house for neighbors to "sit up" with, sort of like a funeral home visit is now. Some didn't have enough money, and the pauper casket was a little ole cheap wooden box. Sometime you could smell the flesh starting to rot.[18] The lodge stopped that, so I reckon the AURS served its purpose.

Endnotes, Forsyth and Atlanta, 1923-1931

1. Founded in 1900, it was named Forsyth Normal and Industrial School in 1902. The Georgia legislature passed an act in 1922 that made it the School of Agriculture and Mechanic Arts for the Training of Negroes. In 1927 it became a junior college. The Georgia legislature again changed the name, in 1931, to the State Teachers and

Agricultural College for Negroes (STAC). It was closed in 1938 when the Georgia legislature transferred its financial support to Fort Valley State College, Fort Valley, Georgia (Source: www.hubbardalumni.org).

2. Mrs. Aquilla Jones Thompson was vocal music teacher at Ballard-Hudson Senior High School, Macon, Georgia. She was Audrey's supervising teacher, winter quarter, 1962, in cooperation with the music department at Fort Valley State College.
3. This spelling not verified. It may be "Maddox" or another spelling.
4. During this recorded session, while talking, he was pointing to a picture of R. J. Moton in *Ebony Magazine*, 2002.
5. This reference is to the professional basketball team, the Atlanta Hawks, member of the National Basketball Association.
6. The letters *ME* was the short form commonly used for Methodist Episcopal.
7. Her sir name may be spelled Rembant or Rembrant. It could not be verified. No given name could be discovered.
8. First Kings 2:1-3, King James Version.
9. Dad recited these scriptures from memory, March 21, 2001.
10. First Samuel 8:4-5, KJV.
11. The news event occurred in 2000.
12. African Methodist Episcopal is a denomination founded by the Black preacher Richard Allen. It was and is operated by African Americans. Morris Brown Academy and Morris Brown College received financial support from the national office of the AME Church.
13. Descriptive of microcultural behavior.
14. January 2, 2003, was the date of the visit to the home of Bessie Ector Rosser in Lone Oak.
15. No verification of spelling of Colly.
16. Minnie Freeman became worthy grand president in 1972 and served until her brief illness and death. She died November 26, 2004. By 2001 there were only three divisions with 350 members. An insurance company had been secured to offer life insurance policies and education plans to remaining AURS members (Source: unpublished document, the American Union Relief Society, 1916-2001).
17. AURS dues were sent to Rosie Passmore in Bremen, Georgia, for 2002, 2003, and 2004. After JR's death, a form to file for his death benefit was sent to Audrey. After several months, a check for three hundred dollars was received.
18. Laws for burial did not include mandatory embalming. These procedures have advanced greatly since the early part of the twentieth century. It has been customary for Blacks to hold a body until all members of the deceased person's family could arrive. Many had to come from other towns and other states. This could take a week or longer before the funeral and burial took place.

HARALSON COUNTY

Year	Age	Education	Children	Births	Events
1931	25				AY 1930-1931
1932	26				AY 1931-1932
					Married March 17
1933	27		John Robert, Jr.	January 3	AY 1932-1933
1934	28	Enters State T & A College, Forsyth, summer			AY 1933-1934
			Samuel Blanton	July 13	
1935	29	Summer school, Forsyth			AY 1934-1935
1936	30	Summer school, Forsyth			AY 1935-1936
			Richard Delano	September 3	
1937	31	Summer school, Forsyth			AY 1936-1937
1938	32		Myrna Suejette	March 9	AY 1937-1938
1939	33		Robert enters grade 1		AY 1938-1939
1940	34				AY 1939-1940
					WWII begins
1941	35				AY 1940-1941
1942	36		Audrey Wynelle	February 18	AY 1941-1942
1943	37	Enters Fort Valley State College, summer			AY 1942-1943
1944	38	Summer school, Ft. Valley			AY 1943-1944
1945	39	Summer school, Ft. Valley			AY 1944-1945
					WWII ends
1946	40	Summer school, Ft. Valley			AY 1945-1946

Chapter 13

BREMEN AND WACO

At Bremen, I taught grades 1 through 7, but a few students wanted to keep coming to school, so I accepted them up to grade 9. I had from one to two students in all of the upper classes. In fact, I saw one of my former students, Ellen Arney, down at Mama's funeral. Ellen's sister Helen Arney was a student in Bremen. She was in the ninth grade, with another girl named Roxie Farmer. I remember her brother, Lewis Farmer, and Henry Lewis, who married Rozie Almon.

Most students only attended their schools until seventh grade, so Waco and Bremen did not have a basketball team that played in the conferences. There was a fellow named Leon Spear who was teaching and had a bus. He drove children from Bremen to Waco. He would haul children all around to play ball, sometime to Tallapoosa. Spear was principal, coach, bus driver, and janitor. As principals, we had to do nearly everything ourselves, including being doctor and nurse.

Bremen, Waco, and Tallapoosa are in Haralson County. The county seat is Buchanan. That's where the school board office was. I had to walk from Waco to get my last check. There was a Black man who was a trusteelike of the school. His boss was the sheriff, Kelly Brown. He told the White man, "If you see a man walking, pick him up and carry him to Buchanan."

Sure enough, a White man picked me up. As it turned out, his brother J. W. White was the superintendent of schools, but I didn't know this at the time. He asked me what I thought about the superintendent. I told him he was a nice guy. I didn't know the man who had given me a ride, but I knew it could be dangerous to me if I said anything bad about one White man to another White man.

Along then they were having a mix-up about Johnson, the Black man who was principal of the school in Tallapoosa. I don't know why he asked me that. Maybe he just wanted one Black man's opinion about another Black man's turmoil. I do know White folk would try to pick any information they could out of you if you were Black and they had the chance to talk with you in private.

We had to take all kind of insults from them. One time there was a teachers' meeting in Cedartown. Pauline's mother was working up there in Buchanan teaching the adult school. I must have been still at the little school in Bremen because we were at the bus station in Bremen. I was in the back, the Colored part. Mrs. Lynch must have been up in the front part.

There was a young fellow driving a Ford car, just a regular five-passenger car, and was hauling folk like a bus. He came from Carrollton and was to pick us up in Bremen, then carry us on to Cedartown. It costs less than a dollar. I heard the man talking, and when I came out, Mrs. Lynch was already getting in the back seat. I was about to get in and sit beside her. There was a White girl waiting to get in too.

He told her, "Get back there with that White woman so that nigger can sit up here with me, so you won't have to sit by a nigger." And that's the way we rode there and back. Mrs. Lynch didn't say anything to me, and I didn't say anything to her until we got to Cedartown and got out of the "bus." Then she said, "Lord, have mercy. I don't know what's going to happen to these people." I think it was Sunday before we had a chance to laugh about it. It was quite an experience.

When I first went to Bremen and Waco to teach, I was boarding with Robert Robinson and his family. I went to all the Black churches around there. I always taught Sunday school and paid in church, the same amount they asked the members to pay. But I never joined a church in that community or any other community where I was teaching. My students belonged to different churches. I didn't want them to think that I thought more of one church than I did of the others.

I did my part as a leading citizen. I wrote a column every two or three months in the *Bremen Gateway,* a triweekly newspaper. "Echoes of the Hills" was a regular feature article.

Chapter 14

MARRIAGE

I met "the girl of my dreams" on Sunday morning, the fourth Sunday in November 1931, at Antioch Baptist Church in Waco. I was standing out in the yard, after Sunday school, waiting for church. And I saw what I thought at the time to be three White girls coming up the road, Pauline, Mary, and Violet. Margaret and Bessie were little girls at that time.

They didn't have any sidewalks. I was standing out near the road. I turned my back to the road because I remembered then you were not supposed to look at a White woman. Black men were not supposed to look at a White woman.

After some time, the girls never did pass. I looked back, and they had turned off at the church. Then, I was puzzled that White folk would come to that Black church. But in a few minutes, after they had talked with the girls that were out there in the yard, about six or seven of them came out to meet me. "I'm so-and-so, and I'm so-and-so." They introduced themselves, and I told them who I was. The others went on off, and I struck a conversation with Pauline. And I didn't leave her anymore.

That Sunday Pauline was invited to dinner with the Robinsons' daughter, who was her friend. They had agreed the Sunday before to have dinner at the place where I was boarding. When Pauline found out I was boarding there, she wanted to break off the dinner engagement. I encouraged her to continue with it, and she did. Pauline went there and ate dinner.

I was talking with her, and I had a pencil and paper, and I wrote, "This twenty-second day in November, I met my wife." She didn't like it! She didn't like my calling her my wife and anything like that. We went back to church that night. They were having night service with kerosene lamps on the shelves on the walls. I stayed with her, and when church was out, I walked home with her, down there where they lived.

When I got ready to leave, I offered to kiss her, and she pushed me back! I got a date set for the next Sunday and went on that way. I never did offer to kiss her anymore until we got engaged. Only when she agreed to marry me did I say, "Maybe we should seal this with a kiss."

We did. That was our first kiss. She remembered that she had pushed me back the first time and said she thought I wouldn't ever offer to kiss her again since she had refused the first time.

I was twenty-four years old, and she was eighteen. She was a pretty woman, and she has been pretty to me all along. She reminded me more of my mother across the forehead than anyone I had ever seen. To tell you the truth, I liked everything about her. She had a pretty handwriting. She was really a good cook because she had been cooking since she was twelve years old.

I liked her and she liked me. She liked me because I didn't drink alcohol, and both of her older brothers did. I was interested in education. She was interested in education. Her dad had been in education. He had been the principal at Flint Ridge, a little community near Waco.

The first valentine card I ever got was from Pauline. It was a picture of a little sea pirate girl with boots on, standing in the sand by a treasure chest. It read, "I chest want to be your valentine." I went to the store and bought her a valentine that said, "If someone wants to be my valentine, I won't stop her." It had a picture of a bottle with a stopper in it. That's how I won over her mother. She thought it was an appropriate valentine for me to send her as a response. We were married by the next time Valentine's Day came.

She wasn't in school because she was at home, helping to rear her siblings, as her father[1] had died three years before. Her oldest brother, Taft, worked for the railroad; he was the main breadwinner. When I asked her mother for permission, I said that we had decided to marry, if she didn't mind. That was the way I asked for her hand.

Her mother said that she didn't want any fooling around. If we meant business, it would be all right with her. But the way she said it, she wanted it to be forever. When the business gets tough, and if we should break up, she wouldn't approve. I remember Pauline telling her, "We can do it. We can stay together."

We talked about it, and it was settled then. After we got engaged, we talked. We talked about different things, how people got along. Some people have to have shrinks.[2] We felt like we could handle our business ourselves. We settled differences of opinion during the time before we were married.

Marriage was supposed to be a lifetime thing. That's what we promised when we married. We talked about the vows that we made. We kept them. We were loyal to each other. We didn't ever have to have a "shrink" come around. We didn't ever have to have anyone tell us how to live together.

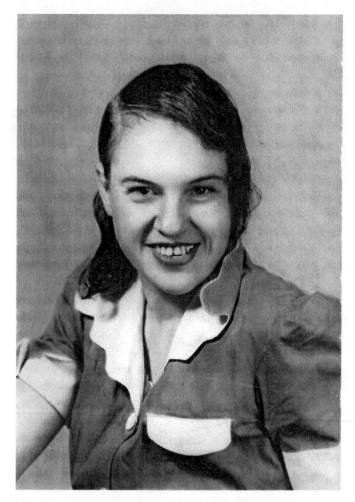

Wife, Pauline Lynch Rosser, Age 29

We were married March 17, 1932, at her home. It had rained that morning, so everything was really wet. My daddy and Momma's cousin came from Newnan. They were my only guests. It was at five o'clock in the afternoon. I was nervous and scared, but I was on cloud nine! I married the girl of my dreams. After that, everybody sat down to eat dinner.

For several months we lived in the house with Pauline's mother and siblings. We slept in the company room.[3] I still roomed with the Robinsons during the week and came back on weekends.

Pauline and I were getting ready to move to Tallapoosa. I helped her brother Tom plow and raise a crop the spring of that year.[4] We planted cotton, corn, potatoes, pumpkins, velvet beans, and almost anything that grows on a farm. Later Tom went to work for a road construction company in LaGrange. When we left, her mother decided to quit farming and just have a garden. Sam was just a boy, too little to be any kind of plow hand. She had those five[5] children to raise, and two were little children.

In August we moved to Tallapoosa to Dr. Wardell's house in Old Town. He was not a medical doctor. He was a preacher and had several degrees.

We didn't want any children. We said that from the beginning until she found out she was pregnant. It seemed like children started coming to our house "two on a mule." Robert and SB were born in the Wardell house.

We later moved to the rented house on Friason Street where Richard was born, then to the house we owned at Forty Connecticut Avenue where Sue and Audrey were born. I don't know if we ever said it that way or not, but we decided all of them would be intelligent and would need a college education.

Mama's health posed a great problem. Five children, born without benefit of a hospital. It was a dangerous, touch-and-go situation. No doctor to help some of the times, but she went through with all that. Her family helped out with the children from time to time.[6]

She would have breakfast and dinner ready and on the table then call me. At first I didn't know how to appreciate all the different things she could cook. Sometime I just wanted milk and corn bread because that was what I was used to. Later, I learned to enjoy and appreciate everything she cooked. She knew how to make a little bit go a long way, and all of it tasted good.

She could can[7] vegetables and fruits. I could grow vegetables and planted one or two apple trees and peach trees. So that's how we made it most of the time.

Once we were about out of food and her folk had kept aplenty stored up, so she sent me with a wheelbarrow to get some food. Now, that was from Tallapoosa to Waco, about seven miles. Going wasn't so bad with an empty wheelbarrow, but coming back was a bit tough.

They filled it with fresh vegetables, some dried apples, a few big glass jars of canned food, and some meat from the smokehouse, as much as could get in the wheelbarrow and stay without falling out.

When I headed back, it was about the time the train went from Waco to Tallapoosa. It costs a dime to ride. I didn't have a dime, even if I could've figured out how to get the wheelbarrow on the train. I watched as the red lights on the tail of the caboose faded out of sight.

It was so dark I could hardly see, but I knew that those tracks would take me home. There wasn't going to be another train to come along until in the morning, so I lit out to walking between the rails. When I got to Tallapoosa, my eyes were used to the dark and I could see how to take my road on home.

That was about the gutsiest thing I ever did, to marry my wife not having any money, not enough education, just an ambition. I didn't get her a diamond ring until twenty years later. It was that long before I could put a ring on her finger. We managed to make it through sixty-eight years and nine months together.

My dad remarried. He and Miss Genie[8] were married in May of 1932, two months after I married. Four months after my first son was born, they had a son named Rofford[9]. He is my half brother. That's why I would say that I had four and a half brothers. Of course, I wouldn't remember anything about Rofford—I wasn't ever around him.

I seem to remember that when Miss Genie died, Pa married, or was going to marry, another lady for a brief while. I don't remember her name, but when she found out my ole man didn't have the kind of money she thought he had, she went on about her way and never came back.

When I think about my ole man, you could say he knew how to survive, body and soul. One thing I know, he wanted to keep something on the *table* all the time. I appreciated that about him. He always had a plenty of food around. He didn't want us to go hungry. That's why he wanted us to learn how to do so many things on the farm.

In the summertime we had to go out and pick peas and shell them. Once while we were shelling peas, he told us we could just boil them in salted water and eat them. And we wouldn't starve. That's probably why I used to plant me a pea patch, especially when I had a growing family.

Another thing I remember. He went to *church* practically every Sunday. One time he had to wear the government-issue, war-surplus boots he had bought for ditching. They were rubber and came up to his knees. It was rainy and muddy, but he went to church. I guess I sort of copied that after him because I always went to church. I have missed more Sundays this year[10] than I had all the rest of my life.

Chapter 15

TALLAPOOSA

I came to Tallapoosa behind slavery time, when they didn't have a high school for Blacks in the area. They tore down a frame building they had for a White school, moved it on the hill across the tracks, and put it back up again. That's what they were having school in. There were two teachers for seven grades. There were 140 children enrolled, but they didn't half come to school.

The principal, a Black man named Johnson, and his wife, a teacher, had worked there for about twenty-two years. The Black people there were sort of like the national parties. If they are Republicans, they have to be against everything the Democrats do, sort of like that. The Blacks who felt they had to be *for* me were against Johnson. And if they were for Johnson, they had to be *against* me. It split some families. Clister Kennedy's for one; his daddy was for Johnson and his mother was for me.

The school board had kicked Johnson out. He had messed up[11] a man's daughter and had to pay another man to marry her. The other man was Dan Louis Thompson. His son was in Robert's class.

Mr. Irby Evans was principal of the White school. I went to him when he was superintendent of schools. He asked me what[12] I had. Was it a degree or what? I told him it was really a high school diploma. Evans hired me. After he met me, we stayed friends until he died. He was against Johnson and felt like he shouldn't be principal after what he had done.

Johnson had been there all that long and had made no progress. I don't know that Johnson had a high school diploma. The only recreational thing he would buy was little red rubber balls. He stayed in town but didn't teach school. His wife[13] taught three or four years with me; then Mrs. Wardell took her place.

The next school year the board elected White teachers and dismissed the board meeting. I went to the mayor and asked him why they didn't elect the Black teachers. There were two of us at that time. That split in the community was still going on when some of the Black folk wanted Johnson to come back instead of me.

Merchants would let him have things, and he ran up a bill. One merchant got on the board of education. He wanted Johnson to work so he could pay on that debt. They put him back in there for that one year. They did a lot of stuff like that.

Reverend Mauldin, a Black preacher, believed the church and school ought to work together. He told a board member, "If you don't put Rosser in there, these niggers goin' raise some hell." The board member said, "They just have to raise hell then because he has another job."

I had got on in the adult school. In the summer of 1933, Huey Dixon and I were discussing jobs the government was creating. President Roosevelt started the FERA.[14] He put up jobs for teachers that were unemployed. It was setting up a school for adults. It was supposed to start with fourteen-year-olds, but many people stopped out of school before they were fourteen.

The federal program was located in the Masonic Hall. That was where I held my classes. I announced my FERA school. Several children who were going over to the school on the hill operated by Johnson quit and wanted to come to my school. I had to question the authorities about that. They told me I couldn't turn anybody down unless they could read. It was in that school that Charlie Byrd learned to read.

The FERA school operated on the regular school year. My check for teaching adult education was more than the state was paying schoolteachers. It paid thirteen dollars and eighty-five cents a week because it was government work. There was regional control of operating expenses. They sent a check every week in the mail. I didn't have to go to work until eleven o'clock. It closed out for the summer months, and it didn't stay but one year.

The next year I wanted the job as principal-teacher. Under Johnson, the school went down so bad until the board came back and got me. Dr. Downey was the head of the board. He raised sand[15] about them running me away by not reappointing me. Dr. Downey gave those White folk such a round for getting rid of me; they hired me and got rid of him.

There was still a tiff among some of the Black folk. They tried to get something up against me. They said, "We don't want ole rascal Rosser. He brought a White woman down here." The superintendent, John White, showed me a letter from the Black citizens. They thought Pauline was a White woman. It said that I was "married to a White woman."

Ole man John White knew Bessie Lynch as a teacher at Waco, and that I had married her daughter. He asked me, "You are married to Pauline Lynch, aren't you?" I said, "Yes." He said, "Well, she told me she was a Negro." He said, "She's white but she's still a nigger." The superintendent was not unfriendly toward me after that. He supported my reappointment. He recommended me for the teaching-principal position.

It was still not settled with some Black folk. When their first scheme didn't work, they tried to accuse me of going with the young girls like Johnson had done. There were two or three illegitimate babies in the community that they tried to pin on me.

Charlie Trailer started a rumor. He said, "Laura Mae said somebody else was that baby's daddy, but he sho' got a head like 'Fessor Rosser." I didn't retaliate because it wasn't true. Time would prove who was right just like it did with Johnson. That baby was named Mary. That Mary grew up to be so much like Johnson she couldn't be denied. Her mother was Carrie Chivers. The baby, Mary Chivers, married Roosevelt Almon when she got grown.

In days to come, I announced at all of the Black churches that school would be opening for the 1934-35 school year. The Sunday before school opened, I stood up in church and announced, "School will open Monday. It's best that you send your children. It doesn't make any difference if I'm the daddy of every baby in town, school is going on, and I'm going to be the principal."

That talk stopped. There was so much turmoil for letting me go and taking back Johnson that for the next fifteen years appointments were routine, and I stayed there as principal-teacher for those fifteen years.

Tallapoosa was an independent school district. There was a glass factory there, when there weren't but one or two in Georgia. That is what brought in money from northern folk coming to settle in Georgia after the Civil War. It was before my time, but that is what sparked the industry.

There weren't many independent schools in the state. After Tallapoosa, Atlanta was the closest independent school, and that was sixty-three miles. Independent schools had their own boards of education and their own superintendent. The counties had to have boards and superintendents.

An independent school was not under the county but dealt with the State Board of Education. It depended on the attitude of the people in the communities. If they decided that they wanted an independent school district and decided that they did not want to be governed by the county, that's the way they got it.

They had enough power to levy taxes and enough finances to operate their schools. They could tax all property within the city limits, for the operation of the city, including the schools, jails, and different departments. They got paid from the county for the students from beyond the city limits who attended the independent schools. The county paid so much, X amount of money, for them to attend independent schools.

A teacher's salary was so low until it took all I had to keep food on the table and clothes on our backs. I had been teaching three or four years in Tallapoosa when they finally gave me that forty dollars a month.

One thing that might be newsworthy happened in February 1935. As near as I can get, it is this: As an independent school district, the Tallapoosa City School Board had spent its entire budget for teachers' salaries for the school year by February. The board had refused to reduce the teachers' salaries as a cost-saving measure. They went broke.

When they got into trouble, all of the members of the Tallapoosa Independent City School Board of Education resigned and left the operation of the schools in the hands of the Haralson County Board of Education. Haralson County agreed with the state to take over and pay teachers for the rest of the school year. A grant from the federal government would be forthcoming.

When they had to get the federal government to pay the teachers, the superintendent tried to get the Black schools closed, but I was hardheaded and wouldn't close.

The superintendent of schools for both Black and White schools was also the principal of the White school. The superintendent ordered me to close the Black school to save money, yet the White school was to remain in operation. He sent that word to the state Department of Negro Education.

They didn't close the White school. I didn't close so they would have to pay Black teachers. The state couldn't run any color line. If you couldn't pay the Blacks, you couldn't pay the Whites. I did not close the school in February.

At the end of February, the superintendent told me that my teachers and I would *not* be paid because he had told me to close the school. He again told me, "You might as well close the school for the remainder of the school year." I didn't. I was not being rebellious. I was trying to educate the Black children of that community.

After three months of no state money, at the end of April, the White teachers had not been paid. A city commissioner named John Heard was dispatched to Atlanta to inquire of the state board why teachers had not been paid since they were to be paid from special federal funds. He made himself known and stated the case. The state board administrator said,

"Your checks are over there."

"Yeah? Why don't you send them?"

"Because you closed the Black school."

"The nigger principal passes my house every day walking to school. He hadn't closed it!"

"Get me an affidavit that it's not closed and will not close until the nine-month term is up and I'll give you your checks."

On Saturday the police brought Mr. Heard to my house, down by the Wardells. He asked me if the Black schools had been closed for the past three months. I told them everything. I told him that the schools had been open.

He had an affidavit from Atlanta to be signed by me. I signed the affidavit that *I had not closed the school and would not do so until the nine-month term had been completed.*

The only way the White teachers got the money, after teaching three months without pay, was because I kept the Black school open for nine months that school year. When Mr. Heard carried that affidavit back to Atlanta, he made several trips because they could not do on the phone then what they can do now.

There were only two Black teachers. He was told, "Their checks are over there too. We can't send out the White teachers' checks and not send out the Black teachers' checks because the federal government has no color line." I remember Mr. Heard telling us that. And he told the White teachers that the only way they could get paid was if the Colored teachers were paid.

We got checks the following week. That is one instance where my being there and taking the action that I took made a difference. Not only did the Black students continue to receive their education, as did the Whites when state allocated funds were exhausted, but White teachers were taught a lesson in civil rights.

The superintendent, John White, was criticized by former school board members for trying to close the Black school, and I was complimented for keeping it open. Some of the White teachers complimented me. One was Miss Heaton, the music teacher. She said, "This is the most money I've ever seen in my life. If you hadn't kept that school opened, we wouldn't have got it."

Many teachers said that they had not had so much money at one time before then. If it hadn't happened the way it did, the White folk would have raised some cane.

In Tallapoosa, the Whites had a music teacher; the Blacks did not. Mr. Heaton was a Jewish merchant. His daughter, named Irma, came and taught us "God Bless America." The board of education asked her to do that. She read music and played a piano.

> God bless America, land that I love,
> Stand beside her and guide her through the night with the light from above.
> From the mountains, to the prairies, to the oceans white with foam,
> God Bless America, my home, sweet home.

She went over that with us several times, and we got to "singing the stew out of it." She could really teach songs. She married a Jewish fellow named Ben Mitnick.

There was a fellow on the new board of education named Crabtree. I wrote in the paper that the board had promised to put a roof on the schoolhouse. I said that we were having such a tough time, our chances were slim to none, so that even a good promise would be encouraging.

Mr. Crabtree said that he could do more than promise. He put a cover on that building that we had over there on the hill. He mentioned that he wanted to show Rosser that they could do something more than *promise*.

The next year I added eighth grade. There were five or six girls and two or three boys. At that time the school wasn't furnishing free books. The board had to put in the application for books. They put in the number of White students and they each got new books. There weren't new books for the Black children. If any White children dropped out, Blacks could get those books for teachers to use.

There were no teacher's editions, and no books had answers. I just had to remember the rule. There was number 13 in the book of exercises in one algebra text. I never did get how to factor that product until the year was gone. The problem was *the product of the sum and difference of two terms*. I have never forgotten how to solve it. The solution equaled *the square of the first term, plus or minus twice the algebraic sum of the first and second, minus the square root of the second term*. Algebra requires common sense.

Ed River's administration as governor established free textbooks. And what gets me is they don't teach all those rules and postulates to everybody anymore. Now they have to go by a computer, and whatever it says is what you have to deal with.

The courses at the Tallapoosa school were patterned after the curriculum at Forsyth. I built that school from ending in seventh grade to being a high school. I added a grade a year until grade eleven was established. It was on a "seven-four" basis, seven years of elementary school and four years of high school.

As I would put on grades of high school, those folk around Bremen and Forty-Eight and Waco would come to that school. About that time, Charlie Dansby had a car and Henry Lewis had a car and one or two others. They rode to Tallapoosa in cars.

My first high school graduating class was four girls. Genolia Leigh, Ellen Arney, Ruby Hinesman, and one whose last name was Crowder were my first graduates. The school had grown from two teachers to five. There was Pauline, Miss McCauley, Rozie Almon, and a fourth was from Newnan.

White folk had baseball teams. One or two of the White boys would help us practice. We had one who could assist with batting and practice in the outfield. He would throw the ball up and knock fly balls to the outfield players. The Hinesmans, Bubba, and his three brothers all played baseball.

Almost every town had a baseball team. That's when I decided to have something new that most towns didn't have. A basketball program. It took on like wildfire!

We did not get paid for coaching. There was no money for baseball or basketball equipment. Basketballs were hard to get. Money was hard to get. We used picks and shovels to dig out the side of the hill to make a level place, a court on which basketball could be played.

In building the field, I had an excellent reason for teaching measurement. We measured out every foot of ground and made the court to official size.

At first, we got an old basketball hull and wet a whole lot of paper and stuffed it into the hull. It wouldn't bounce, but we could pass it to each other. That's how I taught them how to pass the ball. You couldn't dream of what a time we had trying to get started.

We finally got a basketball. Because there was no fence around the court, the ball would take off down the hill toward the Bonner house. When we would play a game, we would have a fellow stashed off down the hill to stop the ball and throw it back in.

I had to build up the program. We didn't get a chance to participate in the tournament until about two years before I left there. After a few years, we had a girls' team, which would beat everybody everywhere. When I got the little girls to wear bloomers and the boys wore shorts, at first there was criticism. Church folk said, "You got them little children out there naked."

Like Charlie Pride says, "I like to think about it, and I might like to visit, but I don't think I could live there anymore!"

I organized the first parents and teachers association, the PTA. There was a big rumor about that because the man who had been principal before me had never heard of a PTA. He called it the TPA.

I got the PTA to sponsor the Boy Scout Troop number 144, but I did most of the work myself. The motto was *Be Prepared.* That was my motto for life because you can't see around the corner until you get to the corner. When you get around the corner, you may meet some kind of traffic jam or something. You have to be prepared for what might occur.

I first learned about the Boy Scouts[16] from a man from Cedartown, Georgia. I have forgotten his name. He brought me some literature about the scouts. I studied the literature, and I liked the oath and motto. The oath was "On my honor I will do my best to do my duty to God, and my country and to obey the Scout Law; to help other people at all times; to keep myself physically strong, mentally awake, and morally straight."

That is what I taught all of the fellows who were scouts under my administration. I didn't know how they advanced from tenderfoot to first class. I didn't know what a coat of honor meant. I just read the book. I took my sons and the Arney boys, Howard "Duke" McDaniel, about ten or twelve boys, and led them all.

Once we had fifteen boys to go camping and stay overnight away from home. The boys had jobs on Saturday. We went out in the woods there in Haralson County, down on the creek, Beech Creek, I think. We had tents and ropes that we tied around trees. And had almost a tree house that went up off the ground at night. At the time I established the troop, I didn't know of any other Black scout troop.

STEWART'S INDIVIDUAL RECORD FOR HIGH SCHOOLS
Tallapoosa High School, Haralson County, Georgia

STUDENTS' NAMES	34-35	35-36	36-37	37-38	38-39	39-40	40-41	41-42	42-43	43-44	44-45	45-46	46-47
Almon, Jeanette													√
Almon, Robert Louis							√	√	√		√		
Arney, Erwin												√	√
Arney, Ruth												√	√
Bonner, Eddie												√	√
Bonner, Ethel											√	√	√
Byrd, Charlsie													√
Byrd, Juanita												?	
Byrd, Mary								√	√	√	√		
Capers, Naomi													√
Chivers, Homer													√
Chivers, Mary										√	√	√	√
Collins, Irma												√	√
Collins, Lloyd D.													√
Conyers, Joyce												√	√
Cox, Zelma												√	
Dunson, Cornelia												√	√
Dunson, Flora										√	√	√	
Dunson, Olus													√
Dunson, Ruth												√	√
Dunson, Virginia													√
Foster, Nathan, Jr.									√	√	√	√	
Hill, Christine								√	√	√	√		
Hindsman, Ruby										√	√	√	√
Kennedy, Christine												√	√
Kennedy, James Clister							√	√	√	√			
Kennedy, Leonard David							√	√	√	√			
Lee, Julia									√	√	√		
Leigh, Clarence											√	√	
Leigh, George							√	√	√	√			
Lofton, Willie Glenn										√	√		
Mitchell, Willie												√	
Reid, Mary Vivian								√	√	√	√		
Scales, Gwendolyn								√	√	√		√	
Sterling, Ora												√	√
Summerlin, David	√	√	√									√	
Thompson, George Samuel										√	√	√	√
Thompson, Henry												√	√
Vaughns, Ethel													√

High school permanent records contained on front side: Pupil's Name, Residence, County and Name of the Superintendent, School year 19:___ Subjects, Grades, grades first and second term, Units earned, and Teachers names. Subjects were grouped: I: English; II: History and Social Studies; III: Mathematics; IV: Science; V: Foreign Language; VI: Vocational; VII: Avocational. On the back of the form: Name, Parent, Residence, Place and Date of birth.

One roster of high school students taught in Tallapoosa

Chapter 16

STRAIGHT AND NARROW

When I was twenty-five years old, I went to register to vote. I was told I had to pay poll tax, a dollar a year for every year I could have voted and didn't. I questioned it.

"What is it for?"

"The privilege to vote!"

"Can you call all those elections back and let me vote in them?"

"No!"

"If not, I don't owe you four dollars."

"I guess you're right." He registered me to vote, and I just paid the dollar for that year.

It was on account of the students that I went to all of the churches in Tallapoosa. There were four Black churches: one had its services the first Sunday, another had its services the second Sunday, another the third Sunday, and the other the fourth Sunday. You would have to walk several miles on Sunday, going to your own church for Sunday school and then to the church that was having services that Sunday.

All of the churches needed Sunday school teachers, so I taught at all of them on their meeting days, the one Sunday a month the church was holding the services. In Mt. Newley Baptist Church, where Mama was buried, was where I taught the first Sunday. The second Sunday was a little Colored Methodist Episcopal Church. There was a trail to it before the Georgia Road 100 was cut. The third Sunday was a little African Methodist Episcopal Church on Brock Street. In Old Town, Mt. Sinai Baptist was the fourth Sunday church. I used to tote[17] my little children away across town to go to Sunday school. We had to walk everywhere.

The churches were in a dispute. I did not get involved in that. The pastor's salary was a quarter. It cost me a dollar a month because I went to four churches and helped pay the pastor's salary in all of them. Class leaders would go around and collect the fee for the preacher's salary and have it ready on the Sunday when the church would hold its service.

There was a preacher named Reverend Eddie Willis, who lived not far from us down on Connecticut Avenue. He was pastor of Mt. Newley and was discussing the pastor's salary situation. I remember quite well because Charlie Byrd and I decided that we would go up to fifty cents instead of a quarter. It was a long time before anybody else followed suit.

I helped quite a few preachers. Reverend Wills, who came to dinner every Sunday at Mary's house in Leeds, Alabama, was trying to build up his sermon about the numbers in Revelation. The numbers used to count the stars in the sky and grains of sand. No man could number them. There used to be a song about that. "My mother is in that number that no man can *number*."

He asked if there were numbers bigger than billions. I began to tell him the Latin prefixes for the numbers: trillion, quadrillion, quintillion, sextillion, septillion, octillion, nonillion, decillion and so on. He said, "Whoo-wee." He later told me that when he started quoting those trillions and quadrillions and the like, he had women swooning all over him.

You might say I was able to hoodwink Mother Nature after I got grown. I was bitten by a black widow spider.[18] That was before we had an indoor toilet. At 44 Connecticut Avenue we had an outhouse. I was sitting over one of the seats doing my business and I went to sleep. When the spider bit me, I woke up. It was dark and I didn't have a flashlight, so I couldn't see what bit me.

When Dr. Cunningham checked me, he okayed it as a bite from a black widow spider because it didn't even swell. When it bit, the poison went on in. He melted some kind of pill in a pan of hot water, took a long needle, suctioned up that water, and gave me a shot in my arm. He told Mama, "If he ain't better by tomorrow afternoon, you better send for me or somebody, 'cause this here bite will kill him."

Mama wasn't teaching, but I was. I was headed to school the next day. I went by Mr. Shell's store. It was in town right across from the railroad station. He said, "Rosser, you're sick. You need to go home." I went back home and stayed two or three days before I was better.

Mr. Shell did call me Mr. Rosser sometime. He wasn't a rebel. He was a liberal. I was in his store one time when some White folk came in and asked him, "Did some niggers come in here the last few minutes?" He told them, "I saw some Colored fellows go by here." Most Whites thought it was all right to call us "niggers," but Mr. Shell was sort of ashamed of it.

I was about forty years old when I first got mumps. Four of my children had mumps. It didn't treat everybody the same. It was harder on you if you were an adult. Some men around there got it. It messed some of them up. It made them sterile. It didn't do anything to me. Robert remembered that he had caught mumps from someone in his class in school, and I may have caught it from him.

My jaw was so swollen on one side I looked lopsided. That side was heavy and seemed like it was pulling me down on one side. Some of the folk who came to see me laughed at me. I guess it looked funny. One of the new teachers, Mozelle Saley, from Cedartown came to see me. She didn't laugh. She said, "I've had it. I know how bad it hurts."

This was when Audrey was a toddler. Mama tied a rag around my jaws some kind of way to ease the pain and swelling. Audrey wanted a rag tied around hers. Mama accommodated her wish, and she walked around or lay down on the bed saying, "I got the mops." She did not catch it until years later when she was a teenager.

I have had a little trouble with city government. The city of Tallapoosa owes me yet for fourteen days' work. When the city went broke in 1935, they were paying me forty dollars a month. They paid me for five days of the school month, and that left fourteen or fifteen days. The county got the money from the state department. I don't know it. I didn't see it, but I heard they paid the White teachers what they were due. But they didn't pay me everything I was due.

I'd ask them about it from time to time. Even after I moved away, I went back. The last time I went to the clerk in the mayor's office, I was told, "We don't deny owing it, but we just don't know who should write the check." His hands were tied.

I guess they had to have some kind of restrictions to save their money. He had to give an account of all the money that went through his hands. There was a debt there before he came in. They had a city clerk that took up tax money and left town with it.

Chapter 17

MORE EDUCATION

I went back down there to Forsyth Junior College[19] for a few years to summer school. I was going to school when my three boys all started to school, before Sue and Audrey were born. I had to go off and leave their mother and leave them. That is the way I got my education. I did my college years in summer school and correspondence courses.

Both Mama and I went back to school during these later Tallapoosa years. It was to stand us in good stead later on. The state was emphasizing college-trained teachers. I finished my undergraduate work at Fort Valley State College in 1950. Mama started her college education there and was to complete hers eight years later. It was tough to do going only during the summers. She had worked for a janitor's salary for several years, at fifty dollars a month.

After all our children had gotten up and some of them had gone, we had one or two stiff discussions. When she went to summer school at Fort Valley State College, one of the doctors got stuck on her. He wrote to her when she came back home. He told her, "Write me if you get a chance." We discussed that. I didn't blame the man from liking her because I was crazy about her myself. We didn't fuss. He didn't write her anymore, and she didn't write him, I don't reckon. If she did, I didn't ever see it, and I didn't think about it anymore.

I called her Pauline until Robert was born. When he was born, I started to call her Mama. And I called her Mama all the time. I called her Mama until she passed this past December. I don't know if any of our children had to have counseling because of it or not. They were all wellborn, and all went to college.

When we got our first house paid for, my wife and I sat down and swore that we would never mortgage it for anything. But when our children got to going to college, I had to mortgage my house in the Northwest Georgia Bank in Tallapoosa. The mortgage stayed there about twelve years. When the quarter or semester would come around, I would have to borrow more money. I was up at Chattooga County Training School when it was finally paid off.

I sent Mamie twenty dollars a month as long as Robert was living there with her. Of course, he didn't know it, or at least I thought he did not.[20] Those Rosser women—Jennie Mae, Elmira, Amy, and Mamie—were pretty sneaky about money.

When I went to Atlanta University, I roomed with Amy. At that time I paid her ten dollars a week for just room, sleeping not eating, during the six-week summer session. I stayed there the next summer, but the third year when I went up there for six weeks, her husband, Jerry Black, told her that I would have to pay to board there if I stayed there any longer.

That was a husband-and-wife situation. He did not know that I had been paying her ten dollars a week for more than two years during the six-week summer sessions. She was working at *Scripto* and making her own money but kept the room money secret from her husband. So I started to leave. She had to admit to him that I had been paying her ten dollars a week the times that I had stayed there for the six-week summer sessions. He then agreed that I could stay there.

That just happened. It was a part of my struggle for getting my education.

Endnotes, Haralson County, 1931-1946

1. Samuel Elisha Lynch, Sr., was born September 5, 1884, and died December 1, 1928.
2. In later reference to the same thing, he used the term *marriage counselors*. The term *shrink* still refers to any psychiatrist and psychologist who provides mental health counseling.
3. According to Bessie Lee Lynch Mitchell Simmons, Pauline's youngest sister, who was seven years old at the time of the wedding, the company room had two beds. JR and Pauline shared one, and she and her sister Margaret shared the other. She and her sister were just children and did not notice anything unusual or different; however, she now notes that sharing the room could have been uncomfortable to the newlyweds (Personal communication, October 2007).
4. March to August of 1932 was the farming season.
5. At the time of her marriage to JR Rosser, Pauline had two older siblings and five younger ones. Her older brothers were William Howard Taft Lynch, age twenty-three (November 11, 1908-March 31, 1980), and Thomas Franklin Lynch, age twenty (March 1, 1911-August 4, 1937). Bertha Pauline Lynch, age eighteen, was born June 29, 1913. Those younger than Pauline were Mary Geneva Lynch, age sixteen (February 12, 1916-July 4, 2003); Ellen Violet Lynch, age thirteen (April 17, 1918-August 4, 1987); Samuel Elisha Lynch, Jr., age twelve (February 15, 1920-July 26, 1998); Margaret Louise Lynch, age nine (May 11, 1922-September 29, 2002); and Bessie Lee Lynch, age seven, who was born August 26, 1924.
6. According to Bessie Simmons, when SB was born, Robert stayed with Pauline's mother and siblings. He liked to stay with them because his grandma would spoil him. She would hold him on her lap and rock him, and he could be the only baby. Pauline's mother was Bessie Lee Stokes Lynch (May 6, 1881-January 14, 1968).
7. Process to preserve food in glass jars.
8. Eugenia Mathis Hutchin was born February 13, 1891, and died January 11, 1941 (Personal communication, July 2002, Ellen Mathis Rosser, widow of William H. Rosser).

9. Rofford Dewitt Rosser, born May 11, 1933.
10. Reference given July 2001.
11. The term *messed up* is used here to indicate pregnancy outside marriage.
12. Education and professional teaching credentials.
13. Mrs. A. L. Johnson taught Robert and SB in first grade, according to JRR, Jr., 2003.
14. Federal Emergency Relief Administration (FERA), 1933-1935.
15. To *raise sand* meant to give verbal protest and complaints.
16. Boy Scouts Oath, Motto, and Law (*The Boy Scout Handbook*, Tenth Edition, The Boy Scouts of America, 1325 West Walnut Hill Lane, Irving, TX 75015-2079, 1990/1910).
17. To *tote* meant "to carry." JRR, Sr., would often carry the youngest child either in his arms or on his shoulders. If he or she were old enough to walk, they were often sleepy or tired.
18. JRR, Jr., says, "I remember when Dad was bitten by the spider. I had to go on my bicycle in the night to Dr. Cunningham's house to get him to come to our house to care for Dad."
19. The State Teachers and Agricultural College, Forsyth, Georgia.
20. JRR, Jr., verifies, "That is so, I didn't know it. As soon as I was able, I began to pay her fifty dollars a month beginning in February 1952. Before that, I gave her some money as I worked part-time at a restaurant in Georgetown, after school, at Miner Teachers College."

CARROLL COUNTY

Year	Age	Education	Events
1946	40	Summer school, Ft. Valley	AY 1946-1947
1947	41	Summer school, Ft. Valley	AY 1947-1948
1948	42	Summer school, Ft. Valley	AY 1948-1949
		Robert graduates high school—eleven grades	
1949	43	Summer School, Ft. Valley	AY 1949-1950
		SB graduates high school—eleven grades	
		Robert enters Miners Teachers College, Washington DC	
1950	44	*Earned BS degree, summer*	

Chapter 18

ALTERNATIVES

The board of education in Tallapoosa met in June and selected the White teachers for the 1946-1947 school year. They announced the appointments in the Bremen paper. At the Black school we had seven teachers. The Tallapoosa board adjourned without appointing those Black teachers. They did not reelect *any* Black teachers to get around dealing with me.

In May of 1946, I had made a request to the Tallapoosa Board of Education for some additions to the school. The board did not reappoint me because I had told them what I was doing in the school, and I needed more space or a new building. I told them what my program deserved, and I wanted something better, a better school building. The school board was short of money, and I was pressing for a building for the Black school.

I had a high school and some graduations for two or three years. The enrollment of 150 students stayed about the same, but with the scouts and basketball, the average daily attendance improved. This increase had held for almost the entire fifteen years that I was there. I couldn't wait for them to act. When the board didn't reelect me, I had to find another job and went on to accept a position elsewhere.

The board was severely criticized for that. Dr. Downey got in behind the board of education for letting me go. He thought I was a valuable man because I was taking those boys into scouting.

Dr. Downey and some others tried to get me back because they missed me so. He had talked with me, and I was almost persuaded to go back. I had a wife and five children to support. I wasn't willing to move them to Carrollton. I was planning to come back when Dr. Downey died. That knocked my going back there in the head.

The Tallapoosa school still had their troubles. A man from Carroll County worked two years, and a woman ran him away. She was teaching on her sister's credentials someway. They both had the same last name. Johnson was still around when I left to go to Carrollton. Last I heard, he and his wife went to Washington DC or someplace where they had relatives.

I had to buy a car to get to Carrollton. It was a green, straight eight, stick shift with a running board and an ornament of an Indian on the hood, supposedly Chief Pontiac. I bought that used 1938 Pontiac in 1946.[1]

They did not require driver's licenses in those early days. I had been driving a truck at Forsyth, from the time they first got a truck, and cars for seven or more years. I chauffeured for the librarian, Alice Carey, at the Cornelia Library on Auburn Avenue. I chauffeured for her for a year and a half.

I went up to Villa Rica to get a driver's license because they were requiring such. The man asked me,

"What kind of car you driving?"

"A straight-eight Pontiac."

He came out and looked at it. "Get in it and drive up yonder and come around over yonder at that corner and come back here, and if it ain't tore up, you get your license."

I tried to teach Mama how to drive. One day when we got back to the yard, she stepped on the gas instead of the brake, and the car hit a column and chairs on the porch. Every time she would get behind the wheel, I think she would think about what happened with the porch. The accident might have injured or killed Audrey and Sue if they had been out there. I believe she couldn't get over that thought. She got a learner's license but never did get her driver's license. I didn't encourage it. I would drive her where she wanted to go.

At that time the people down in Carrollton were trying to get their school accredited, and they didn't have a librarian with any training. A library was one of the requirements for accreditation of high schools. L. S. Molette was principal at the Carroll County Training School in Carrollton. Like other Negro schools, it had grades 1 through 11.

He was trying to get the high school accredited and needed a library science teacher. Since I had library science training, Molette asked me to apply for a job at that school. That is why he recommended me. I received a contract from the board, and I signed it. So that is why I accepted the teaching job with them and functioned also as assistant principal. I commuted back and forth to Carrollton, twenty-one miles one way for the next four[2] years.

The second year they needed a seventh-grade teacher at the school. Mama applied and got that position. I took my children out of the Tallapoosa school, and all seven of us rode back and forth. It was tough. It was long days, driving there, teaching all day, then driving back.

Mama never complained, but she had just as much work to do as I did. We had to get up at five in the morning. Bless her heart, she cooked a good hot breakfast for us. She had to get clothes ready for everybody, and she had to plait our two girls' long hair.

We had to leave real early, and we got back late when we had to stay after school for teachers' meetings. Sometime we had to go back at night to a PTA meeting or basketball games. Sometime it didn't pay to drive home and turn around and come back.

One time we had a night meeting. All five of our children went to a movie. I think it was *Gunga Din*. The meeting lasted until nine thirty. By the time we picked them up, they had seen it four or five times.

Mama was in charge of collecting money, the admission at the door, when we had basketball games. I think they paid twenty-five[3] cents to see a game. Well, she'd take up the money; then she had to count it and get it ready for the bank and lock it up. She was such an honest person she could be trusted with that.

What Mama and I had to do didn't stop when we got home. We had lesson plans, and we had our own reading and writing and assignments to do. You see, we were both taking correspondence courses at the same time to try to get our college education. It sure enough was a "tough row to hoe."

Sometime when we'd go back and forth to Carrollton, we had a little humor. I remember when Audrey saw that sign that read Watch for Trucks. After a while, she said, "They need to take that sign down, I've been watching all this time, and I haven't seen a truck yet." Hahaha. We teased her about it. Of course, she was just six years old, but she was in the third grade, and she could read.

I guess the worse thing that happened was when I got run off the road by that ole eighteen-wheeler. We all remember that. The White fellow that was driving was going real slowly, and it wasn't uphill. I learned afterwards that he just did that to antagonize me. It was just our two vehicles on the road. I needed to get on to school, so I pulled over in the left lane to pass. Just then, he pulled his truck over in that lane to cut me off.

To avoid smashing into him, I steered the car over to the ditch off the highway. I saw him lean out of his window and laugh as he drove on off. We were not hurt. It just shook us up. I think SB's head hit the top of the car, and he had a bump on it. I got the car back on the highway and went on to school. That's what I had to do.

Chapter 19

A FULL LOAD

At the Carroll County Training School, I took on a full load. I taught some math and all the history: Georgia history, American history, and Negro history. I took care of the library, was the assistant principal, and did some coaching after school. I had been a working principal, not just a supervising principal, because I was in a small school with ten or fewer teachers. With just ten teachers, a principal had to teach some courses himself and herself. I had taught mathematics throughout my career. I was used to being principal, teacher, and coach.

I got the library in good shape. The idea of the Dewey Decimal System helped to keep account of books. We had a book pocket to keep account of it and how long it was out and when it was due back. Every book had a discrete number according to categories. Before that system, we had no organization. The reference books, stories, and every kind of book were arranged by authors' names. Whew!

The high school courses at Carrollton, like those at Tallapoosa, were patterned after the curriculum at Forsyth. They taught geometry, algebra 1 and algebra 2, history, science, English, French, home economics for the girls, and agriculture for the boys. Agriculture classes were later called "shop" when they dealt with building things out of wood.

The training school was the only high school in the county for the Black children. I had to come up with what needed to be done to get the curriculum[4] and instruction in shape because I had to write the appeal for accreditation.

There were some good teachers there, and some were dragging the program down. Molette was playing favoritism and nepotism. I talked with him about the departments. He was a sort of buddy with Crogman Mullins. Crogman was in the Ag[5] department and coached a softball team along with Leroy Childs. Molette hired three women in Crogman's family that were teachers, two daughters and a sister. One of his daughters was the home ec[6] teacher. I wrote that all she did was make salad. She didn't know how to cook anything. I meant it as a criticism. I would have removed her, but I couldn't go over Molette's head.

In general, we were trying to build up the school program. He had a pretty good scouting program. The big problem we had was keeping children in school. It was

101

farming country. It was a problem for a long time with adults, White and Black. They'd stop their children out to pick cotton or do anything. There were some parents who like my momma wanted their children to go to school. And there were some who wanted their children to work. The way we got money for teachers, equipment, and so forth was by attendance.

Basketball had been a drawing card for children staying in school in Tallapoosa. I thought it might work there too. Molette agreed. He knew I coached winning teams. When I was in Tallapoosa, I had to go down there several times and drill on boys' and girls' basketball. He knew people in the community would pay to come see winning teams play. That would mean money for the school program.

Teachers were not paid for coaching. They volunteered their time to coach in order to have teams of girls and boys. Many times, the principal had to do the coaching. At all of the other schools in which I worked, I was also the coach. I did not have to do all of the coaching at Carrollton. I was only supposed to be in charge of the boys' team.

The assistant coach was that same Mullins woman that taught home ec. She had been in charge of the girls. She never even got any tennis shoes. She would get out there to practice in her dress clothes and have them shoot free throws ten or twelve times. Sometime the girls would miss all of them. Lord, Louella![7]

It was a problem for a while, but I ended up coaching the boys and girls. They had rules for girls that were different from boys. Six girls in play, with three each, playing the defense and offense on different halves of the court. Girls could dribble only once before they would shoot or pass the ball. Guards could not cross the centerline on defense, and forwards could not cross the centerline.

Later when the rules for both Black and White teams were changed and girls could cross the centerline the same as boys, it was easier. I practiced my girls with the boys' rules before the rules were the same. That is why they could beat most of the girls' teams they faced. I had a girls' team that was ready for the conference before the boys' team was.

John Anderson, Tammy Thomas' brother, was principal in Mt. Lowell. He'd let me have his car to go home after practice when I was having car trouble.

I had the championship team for a long time. I picked out a team that was the best and called it the "varsity" team. The other players were the "rookies." I would play with the rookies. When we started the season, the rookies could win. The varsity played the rookies until the rookies could no longer win. Then the team was ready for championship basketball; they were ready for conference games. The conference had a committee to schedule all the games.

It was always tough to find a basketball. But I trained my girls and boys with a cover stuffed with wet newspaper. It would not bounce, but they learned to pass, catch, and shoot with it. Bouncing the ball on the rocky ground was not effective anyway. This was especially so when we had to play on a dirt court.

Molette didn't play sports. He was a preacher. He pastored churches while he was a principal. When we got a teacher that could coach football, then that started the football team. I think Childs tried it for a while.

Molette owned the only gym around. But it was not on campus. He used to let the Carroll County Training School practice there coming up to a tournament. When we had tournaments, you had to sing his song. We had to use his gym. People paid him to let the schools have tournaments in his gym.

He liked money. He had a dwelling in Fort Valley. I stayed in it one summer. I paid rent to the woman who had charge of it. I don't know what happened to his property, and I don't know what happened to him.[8]

He was light skinned. He had help from one of his White ancestors. He had a nub in the place of one of his hands, but he drove a car. In my early teaching there, he didn't have a wife. Leroy Childs and I had "white" wives. Molette wanted them. They both taught in the elementary department there.

Somewhere in the back of my mind, he approached Childs's wife[9] or my wife. Childs and I knew we both had good-looking women as wives. And we couldn't blame a fellow for approaching them if he thought he had a chance. That is the reason why we didn't kill him.

I came there because he had recommended me for the job. In the last part, it had a bitter end. The superintendent at Carrollton called me in. He told me I had two choices: I could resign or he would have to fire me because Molette said he could work with me, but he could *not* recommend me another year. So I agreed to just resign from that school.

They could write it down that I resigned; however, I was actually fired. That is how I got the idea that Molette had approached my wife, and he was scared she would tell me about it. He was actually scared of me on that account because I didn't mess around.

I never did explain all of this to Mama.[10] She was brilliant in the circumstances in which she lived. I was pretty hotheaded when it came to my children. But she kept me from acting on my first impulse. Problems with our children first started when Robert, my oldest boy, finished high school.

He applied at Morehouse. They had an application they sent out. At that time, they had an entrance test with certain information they wanted on it. It was before the SAT and ACT were made popular. It was sent to Molette.

Molette administered it to him. Robert filled it out, and I don't know what he put on it. I never saw it. He answered the questions the way he wanted to or understood.[11] Based on that, the folk at Morehouse turned him down. The registrar, named Whitaker, was the one who wrote the letter. In fact, they sent him a letter and told him that he didn't seem to be the caliber of student they wanted and advised him to look for vocational school.

I was going to write them a letter sailing them out about not accepting my boy. Mama advised me not to do that. She saved me a lot of embarrassment that I could have brought on myself. I wanted to bring it on myself. I was crazy enough to get "*naked*" about it when anybody was doing anything to my children.

Up to that time, I had high regards for Morehouse. I lost all faith in the college when they told my boy he was not the caliber of student they wanted. I thought about it again when Robert got his doctorate degree.[12]

Endnotes, Carroll County, 1946-1950

1. No civilian automobiles were manufactured between 1942 and 1945 because of World War II.
2. At the beginning of the 1946-47 school term, the sons were ages thirteen, twelve, and ten. The personal notes that follow express what they remember. The daughters were ages eight and four.
   - "Only Robert and SB, the two older sons, went with Dad to Carrollton the first year. I had to stay to help my younger sisters get to and from school in Tallapoosa" (Personal note, Richard D. Rosser, December 24, 2006).
   - "The first year Dad carried Irvin Arney, James Lee, Howard "Duke" McDaniel, Howard Vaughn, SB, and me with him to Carrollton. All of us except SB were his boys' basketball team from Tallapoosa High School. The five of us were all in the 10th or the 11th grade and SB was in the 9th grade. The next year those boys didn't go with us. Instead our whole family went" (Personal note, JR Rosser, Jr., March 13, 2007).
   - "I vaguely remember the first year we went to Carrollton. I was on the basketball team; but, I stayed on the bench. I did not play in a game" (Personal note, Samuel B. Rosser, MD, March 16, 2007).
3. "I remember the costs of admission to the basketball games were twenty-five cents for adults, most of whom were from the community and the county; and, ten cents for children and students" (Personal note, Myrna Suejette Rosser Riley, March 13, 2007).
4. "The courses were somewhat disorganized. SB was a year behind me, but he and I were in the same *English Literature* class. I remember it because I got a 'C' and he got an 'A.' I don't remember what curriculum it was, but Molette taught me two courses. One was *Personal Problems* which was about choosing a mate and marriage. The other was *General Business,* which included how to write a check and how to balance a checkbook" (Personal communication, JR Rosser, Jr., 2003).
5. Agriculture as a field of study was truncated to ag.
6. Home *ec* was short for home economics. Instruction in both home economics and agriculture were commonly found in the curriculum of Negro training schools. They were taught on beginner, middle, and advanced-skill levels with pupils being required to take a class in ninth, tenth, eleventh, and twelfth grades. A school club for pupils in those classes that existed on local, regional, and state levels was the Future Farmers of America for boys and the Future Homemakers of America for girls. There were the same classes

for White pupils in rural schools across the South. To make a distinction, the clubs for White children were called the *New* Farmers of America and the *New* Homemakers of America. In the 1950s classes in languages such as French and Spanish were approved as electives for Black boys and girls who were going on to colleges and universities for further study.

7. An idiomatic expression that meant the same as "My oh my!"
8. Lemuel S. Molette died January 23, 1983 (Source: West Georgia Memorial Library, Carrollton, Georgia). JR Rosser, Jr., remembered that Molette's middle name was Scott.
9. Mrs. Vivian Childs, wife of Leroy Childs, is deceased (Personal communication, Ida Tarver Jones, October 2007).
10. Pauline Rosser said, "The students at Carroll County Training School adored JR, while they were not fond of Molette" (Personal communication, Myrna Suejette Rosser Riley, March 2007).
11. "I recall taking an exam of some type. I do not recall the nature of the exam, but I was not interested in going to college. I planned to join the Navy and get away from home and earn some money to help my family with expenses for the other children" (Note by J. Robert Rosser, Jr., August 5, 2001).
12. J. Robert Rosser, Jr., earned a doctor of education degree from George Washington University, Washington DC, in 1978.

Serial No. 12847

Race Negro

State of Georgia

Department of Education
This is to certify that

JOHN ROBERT ROSSER

having furnished the State Board of Education satisfactory evidence of good moral character and scholastic training, a Bachelor's degree from a standard four-year college with minimum credit of eighteen semester hours in Education, as specified for Georgia teachers, and successful teaching experience of forty-nine months, is hereby granted a

LIFE PROFESSIONAL ELEMENTARY TEACHERS CERTIFICATE
FOUR-YEAR COLLEGE

with permission to teach Elementary Grades and Teacher-Librarian

When this certificate is countersigned and approved by the State Superintendent of Schools, the holder is duly authorized by the State Board of Education to teach in the public elementary schools of Georgia. This certificate shall be valid for life, unless the holder fails to teach for a period of five consecutive years, whereupon it shall automatically lapse. A certificate that has lapsed may be reinstated upon satisfactory completion of six semester hours of professional training at a standard institution.

Given under our hand and in accordance with the rules of the State Board of Education this 1st day of July nineteen hundred and fifty.

Governor and Chairman
State Board of Education

State Superintendent of Schools

Teaching Certificate, 1950

Serial No. 12847

Race Colored

State of Georgia

Department of Education

This is to Certify that

John Robert Rosser

having furnished the State Board of Education satisfactory evidence of good moral character and scholastic training, having had three years of successful teaching experience, having met the requirements for the Professional Teacher's Certificate based upon four years of standard college work and having in addition ten semester hours of approved preparation for the work of the principal, is hereby granted a

PROVISIONAL PRINCIPAL'S CERTIFICATE (P-4)

When this certificate is countersigned and approved by the State Superintendent of Schools, the holder is duly authorized by the State Board of Education to serve as principal in the public schools of Georgia for a period of three years from date.

Given under our hand and in accordance with the rules of the State Board of Education this **first** day of **July** nineteen hundred and **fifty-one**.

M. D. Collins
State Superintendent of Schools

L. M. Lester
Director of Teacher Education and Certification

Provisional Principal's Certificate, 1951

CHATTOOGA COUNTY

Year	Age	Education	Events
1951	45	*Enters Atlanta University, summer* Samuel enters Clark College, Atlanta	AY 1951-1952
1952	46	Summer school, Atlanta University Pauline, summer school, Ft. Valley	AY 1952-1953
1953	47	Summer school, Atlanta University Pauline, summer school, Ft. Valley Richard graduates high school, twelve grades	AY 1953-1954
1954	48	Samuel earns BS degree Richard enters North Carolina A & T University, Greensboro, North Carolina Robert drafted in US Army	AY 1954-1955 Brown versus Topeka
1955	49	*Received P-5 Certificate* Pauline, summer school, Ft. Valley Suejette graduates High School—twelve grades Richard joins army	AY 1955-1956
1956	50	Pauline, summer school, Ft. Valley Suejette enters Fort Valley State College	AY 1956-1957
1957	51	Pauline, summer school, Ft. Valley	AY 1957-1958
1958	52	Pauline earns BS degree Audrey graduates high school—twelve grades Robert earned BS degree from DC Teachers College	

Chapter 20

MOVING AHEAD

When I graduated from Fort Valley State College in August 1950, a person who had taught me and a person I had taught graduated in that same class. The one who taught me was Nellie Thrash. The one I taught was Jesse Almon. I gave him his high school diploma from the eleventh grade at the Tallapoosa school. The first "daughter" from the Tallapoosa school that earned a college degree was Rozie Almon Lewis.

At that time, the directors of Negro Education were White. It was a separate division for the State Department of Education in Atlanta. Our director was named R. L. Cousins. He knew me personally and knew Mama personally.

While I was in summer school, I went to his office. I told him I was out of a job, and "I have a wife and five children, so I have to work." He knew we were living in Tallapoosa. He said, "Would you like to go farther up north in the state?" I said, "You know I have to work, I will work anywhere." He picked up the telephone and called C. E. Atkins, superintendent of Chattooga County schools.

Mr. Cousins got me an appointment. He asked me, "Could you go to Summerville?" I said, "Yes, I can, it is not so far from where I live." So he asked Mr. Atkins when he could talk with me. He told him, "Saturday, if he wants to." I went back home and went to the superintendent's office up there. That is where I met Professor Holloway. The Nichols family were big shots in that school.

I became principal of the Chattooga County Training School in Chattooga County. The same Rembert who sent me to Morris Brown, her son had been teaching at that school, and they had run him out from up there. I don't know whether he left or was fired, but he did not want to go back. He had a program that there was a charge to attend and wouldn't let some of them come to it, so Alvin Evans and some of them threw rocks through the windows.

There was an enrollment of about three hundred during my tenure there. I had about twelve teachers. Mama was one of them, and one of the best. She taught fifth grade, then seventh grade. Then when I needed her to, she taught first and second grades, together in the same little room.

After teachers had two years of college, they could get a certificate. A lot of them got that and quit. I had a teacher that taught with me at Lyerly, in the little school in Holland. She had two years of college training, and she never did get any more. I couldn't fire her because her mother cooked for the sheriff.

I was certified and had finished college. I had done work on a master's but had not finished it. I had a P-Four certificate.[1] For a P-Four certificate, the principal would get five dollars per month for each teacher supervised. At P-Five the rate was fifteen dollars per teacher per month.

The State Department sent out notice of an application for a P-Five certificate. They sent me one. Of course, I was working on my masters at the time. The conditions were that a person need not apply if they did not have at least twenty years' experience. I had more than twenty years' experience as a teaching principal.

The superintendent was new and did not know anything about the announcement. He had to recommend any applicant from his county. I filled out an application, and he signed it and sent it in. The next thing that I knew the state board sent me a P-Five certificate, and I worked at that level until retirement.

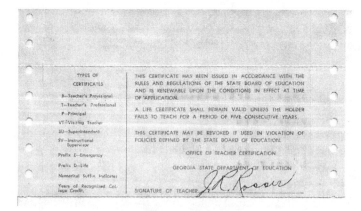

P-5 Certificate, front

P-5 Certificate, back

We finally got teachers' salaries equal. When they had to have *so many*[2] teachers for little children according to ADA, average daily attendance, you could do all right as long as the count was by percentages. For example, two children in a family, if one went to school, that was fifty percent. That sounded good—50 percent of the children were in school. But when they had to count the children, *one* did not count up much. Pay was based on the number of students enrolled and the average daily attendance of the enrolled students.

It was a rural county, a cotton-growing community. It was funny that the dark people, Negroes, would kill each other, one or two a year, in that community. The whole county changed under my leadership at the school.

In a little whistle-stop called Holland, Mr. Ratliff ran the store. White men in the community used to sit around there and talk. One of them I knew about was younger than me, a lawyer named Bobby Lee Cook who got them off on technicalities, including confusing witnesses to shootings: "You saw the smoke, but did you see the shot?"

After I had been there about five or six years, I went in there and Bobby Lee Cook said, "I used to free those niggers that killed each other. Ain't had a nigger killin'" since you been here." A White fellow who ran the store, who was very profane, said, "Rosser, we used to have a nigger killing about twice a year, and since you've been here, they ain't had none. What did you do?"

"I told them there wasn't any use of it. And I had an audience. I showed them that I could shoot, but I didn't shoot anybody. I told them that it was not worth it, *killing a man*." So the White man who ran the post office said, "I knew damn well you did something about it." I didn't explain anything I had really done. I don't remember him ever talking to me after that.

The first year I was there, I went hunting with a fellow and made one or two of what they called "impossible shots." They had a class for veterans on this program that the government set up in schools and were having night schools for the men to go to. It happened one evening that I was looking out of my office window and heard the dogs chasing a rabbit across there, and I saw him run across the street and down to the woods.

I got my shotgun and got out there, and about that time, they were turning out the school. The veterans called the dogs and said, "He went down that way and he'll come back." After a while, we saw the rabbit under a small pine tree. There was no undergrowth under the pines. We saw the rabbit tipping around and met up with an old slow dog. Then he cut out in a hurry. I was standing up there with my shotgun. The rabbit started up, and the dog got after him, and he tore out up close to where we were standing, and I shot him in the right breast. I shot the rabbit and he jumped up high.

And the fellow told me, "That was a pretty good shot you made there!" And I told them, when I had a crowd there, "That's the time to shoot, when you see a rabbit or squirrel or something like that. You all have been up here shooting each other. You've got to stop that." There were two among them who had been involved in shootings of others.

A friend who had killed somebody told me that every time he closed his eyes, he saw that fellow dropping down on his knees with his hands up hollering, "Please don't shoot me, please don't shoot me." And he went on and shot the man.

Later I went off hunting with a bunch of Whites and Blacks, and we didn't have any luck until finally the dog treed a squirrel and chased him out to the edge of the woods, and he went up a tree. And the dog treed him under a pine standing about ten feet away from the muscadine vines in the corner of the woods.

A White fellow had to catch all of the dogs and hold them. They told me, "If he jumps, you shoot him." He went around to the other side of the tree. The squirrel jumped for the muscadine vines, and I knocked him out of the air like skeet shooting. And that put the Whites on to my being able to shoot.

They treated me like a "king" up there because I had an audience when I did some shooting. The truth of what caused them to treat me like a king was just because I could *shoot*. I've thought about it since then. What if I had missed that rabbit and that squirrel? They would not have had any respect for me. They would be laughing yet that I had missed.

When I went to Lyerly, no Black person had a telephone. The lines passed their homes. Black folk owned a lot of land up there. The first year I applied, the folk said it would be too expensive. But the telephone company had posts going across Black folks' land. Most of the Black men were Masons.[3] We had a strong Masonic organization. We got together and threatened to cut the posts down that were on Black folks' land if they didn't put us a phone in that school. I got a phone put into the school. So that is how we got the phone that students used in home economics.

I established a Ground Observers Corps site for the United States Department of Defense. The job was to identify planes from sight, before radar was established, and call in the description and direction they were going and so forth. Our code name was Echo-Echo three-two red. I was the only Black supervisor. The United States Air Force awarded me metal bars and a certificate for one thousand hours of service.

Ground Observers, Certificate, United States Air Force

Chapter 21

SIGNS OF THE TIMES

There was total resistance to integration in Georgia. When I went to Carrollton and to Lyerly, this was evident. Carrollton bused students from all over the county to Carroll County Training School. Schools were fully segregated in both those places. There weren't any Blacks going to White schools and vice versa.

We had tried to get a Black football team. We had two high schools, one in Summerville and one in Holland. Neither one had a football field. But we had some boys who ought to have the opportunity to play football. So the White coach at Summerville helped us with it. He went to work with them. Some of the boys from my school and some from the other school got together and practiced.

When we got ready to play a game, there was only one field fit to play on. That was at the White school. Aycock[4] in Rome was going to play the Summerville-Holland team. They were going to have a game at the White school's field. When they got ready to play, Marvin Griffin was governor of Georgia. He heard about it and sent his troops up there and torpedoed the whole situation.

We had advertised the game, but when it was time to play, even with all of the training and preparation we were prevented from playing on the field. Marvin Griffin said, "It violates our principles of segregation." Just for playing on the field where the Whites play! They were not going to be playing a White team, but a Black team. It didn't make sense. Now Blacks and Whites play together on the same team in high school, in college, and in the pros.

I wrote about it in the *Atlanta Constitution*. Its editor was Ralph McGill. I wrote a letter and he printed it. I said, "What are their principles of segregation? They might as well get two electric chairs because of this principle. Were they going to have two electric chairs? If they didn't have two and a Black man was sentenced to the electric chair and was electrocuted in the same chair that a White man was, would it violate their principles of segregation?" I said, "Blacks were fried in the electric chair where Whites were fried, *it* violated their principles of segregation."

He published two or three of my letters. Race was not considered. It might be that a lot of them did not know that I was Black, just thought I was a White guy who sympathized with the Blacks. The last one was an article on the closing of the training school.

They moved all of my high school students to Summerville. It was costing the Chattooga County Board of Education too much money to bus students. There was a strong affinity for large high schools. They thought the bigger the school was, the better they could teach.

There were a whole lot of Black folk who lived on the mountain. Two buses came to Holland from over the mountain for high school and grades 1 through 8. Gore[5] was a shorter route to Summerville, on Highway 27.

We had one or two problems about attendance. It was a cotton-farming community. There was a White man that owned much of the land and had people living on it. Cotton was the only thing that brought in any money there. Students stayed out of school for picking cotton.

A bigger reason they didn't come to school was that the road on the mountain would stay impassable for months at a time during the winter. The winters were pretty severe. When it would freeze, no sun could get in there under the pine trees, and the road would stay frozen for months.

We had a director named L. C. Harper. He was a principal from Booker T. Washington High School in Atlanta. He came to Lyerly and told them if they have an attendance officer, "all these little Black children will be coming out of these mountains." He talked about having an attendance law to apply to Blacks, and others, in the mountains. It was a long time before the attendance law applied to people in the mountains.

Ultimately, all Black high school students in the county were bused to Summerville. My baby daughter had to ride the bus, pass two White high schools, to get there. They hauled Black students to the school for a long time to avoid integration.

Dewey D. White was principal of the Black school at Summerville. White had lived there ever since he came over there from South Carolina. I think he had earned a master's degree. The last I heard from him, he was in a little school out from Macon.[6]

Mr. Atkins left the superintendent's office the year I was employed. Mr. Lowell Hicks replaced him. Hicks lived in Chattooga County. The why, I don't know exactly, but when Hicks became superintendent, he catered to White. Two years before moving my high school students to White's school, my daily attendance earned me another teacher. But he gave the other teacher to White.

The Chattooga County Training School was "torn up" in 1957. It was cut off from being a "training school" and became an elementary school. I remained principal there one more year. After I left, I was told by some of my friends in Lyerly that the school went down. I never understood just how it ended.

Chapter 22

GEORGIA INTERSCHOLASTIC ASSOCIATION

Before integration, we had an organization called the GIA, the Georgia Interscholastic Association. We had *classes* of schools according to enrollment in high schools. We had class A, class B, and class C high school athletics. The smallest schools were class C; the middle were class B, with at least 250 enrolled; and class A were the largest schools. Part of the time I was in a class B school and most of the time I was in a class C.

The GIA handled track and field, basketball, football, and baseball. My schools had basketball, no football, and very little track and field. The athletic program was how we encouraged Black students to stay in high school long enough to graduate. We used basketball and football to keep children in school long enough to have a program and long enough for them to get a diploma.

I was in the leadership structure. Our schools would have regional directors to handle the tournaments. When it came my turn to serve as tournament director, I had a dominant team in our region. The school for the "deaf and dumb" in Cave Springs had a pretty good little team. At tournament time, the strongest team would usually be seeded against the weakest team.

I had a problem coming up there for me because I had the strongest team in the region. The school for the "deaf and dumb" had the weakest team in the region. The director of the deaf school came to me and told me, "I want my team to have a chance in the tournament, and I don't want them to have to meet your team to start off with."

That was the big problem that I faced. It was from a sympathetic standpoint. I felt sorry for the "deaf and dumb" children. So I worked around and got them to meet another team. I had to do a little politicking to do it, but I did it. They didn't have to meet my team. They won one game. The Chattooga County Training School team won the tournament in 1955 and 1956.

My biggest work in the tournament business was while I was at Chattooga County Training School. The team that had dominated the C class was at Calhoun; the principal was a woman named Betty Smith. After she beat them a couple of years, the team at Dalton had jumped from class C to class B to keep from having to play her team. He did not have as large an enrollment as Smith had, but he went up to class B.

They would let you go up and down[7] if you wanted to. Cartersville was in class B. Hightower was up in Dalton. I think he went to class A and stayed up there several years, but finally he had to drop back to class C. It had almost come around to my term again, and they integrated the schools.

When integration came, the superintendent and board of education made the decision, and we had to follow suit. They cut out the Georgia Teachers and Educators Association that governed the GIA, so all of our organizations ended. Several schools dropped out. Our little school just about played out because there were so many larger White schools in the region. We had to join in other things[8] with the White teachers association.

GEORGIA INTERSCHOLASTIC ASSOCIATION

OFFICIAL PASS
1956 - 1957

PRINCIPAL John R. Rosser

COACH

L. M. TAYLOR,
EXECUTIVE SECRETARY-TREASURER

Georgia Interscholastic Association, Offical Pass

Chapter 23

UNWRITTEN RULES

White men held the power in these small towns. I didn't have the opportunity to run for senator or mayor or anything. The first Black I knew of elected to anything was in Hopson City, Alabama. The reason was that only Black people lived there. That's the only place I ever threw a bowling ball.

I still had the 1938 Pontiac. Pauline and I went to Anniston, Alabama, to see her sister Bessie. Jake was too "high falluted" to hang around with me. While Mama was visiting Bessie, I went with Chillie to the bowling alley. It was the only bowling alley I knew where a Colored person could go. Not in Atlanta, that I knew of, and not in Forsyth.

My first ball went into a gutter, and that was my dime gone. I never tried to bowl again. I never went to another bowling alley. After I got on up, I didn't have time to go to another one. It wasn't as popular then as it got to be, and it's not as popular now as it once was.

If you were a Black man driving a nice-looking car, you caught the attention of policemen, sheriffs, and the state patrol. They felt like you ought not to be driving a nice car, and they felt like they ought to do something to try to show authority over you; but they didn't always know what to do, especially if you weren't breaking the law.

Once in the late 1950s when I carried Richard and Irma to the airport in Rome, a fellow in a sheriff's car stopped me. He had stopped the man in front of me. He came over to the car, asked to see my license, and then said, "I'm trying to 'ketch' somebody hauling liquor." He made out he was giving me a ticket so he could give the White man a ticket. He said, "I know you boys ain't hauling no liquor, and it ain't no use of me giving you no ticket, but I got to give this fellow one, and I'm just acting like I'm givin' you one too." He handed me a blank ticket and motioned me to drive on, so I did.

Another time a sheriff stopped me out on a little road. He said I passed him when the yellow line was on my side of the centerline. I said, "I don't know if it was a yellow line or not, I know the road was straight, so I went on by." I had to go back to LaFayette for him to write me a ticket. He told me I could come there and meet court.

I asked, "For what purpose?"

"Don't you want to go to court?"

"No. Ain't no use of me coming back. What do you have as charges?"

"Twenty-five dollars."

"I'll just write you a check for it. What would my word be against yours?" My word wouldn't be a fart in a shit shower against his. It wasn't that I had done anything wrong. I had a new 1954 Pontiac and my tag was J-15. I guess they wanted to bully me because of that.

We could buy gas, but we weren't allowed to use the restroom. We would have to stop by the side of the road before we got to a little town where we could buy gas. Along the routes that we traveled, we knew the places where it was thick enough to hide you from the road, but clean enough to squat and do your business.

And it had to be that way on both sides of the road. There weren't any interstates[9] then, and you could get to the other side without too much trouble. Any of the boys that had to go would go with me on one side of the road, and Mama and the girls would go to the other side. That's how we had to do that, to be excused, to tend to functions of nature.

The country churches had outhouses, but we didn't stop and use them because it would be trespassing. So we'd stop in the woods. It wasn't against the law because animals can pee anywhere. I knew all the places between my home and Savannah.

When we were living at Lyerly, I remember Mama, Hermanita, and me going from Leeds, Alabama, to Columbus. We went to Leeds to see Mary and Chillie. Then, we left Leeds going to see Margaret. We entered Georgia through Phenix City and went on to Columbus. We had to get gas, and the ladies needed to use the restroom. Two boys who ran the station were members of the Auburn University football team. We asked them, and they said, "Yes, you can use the restroom."

I thanked them and Mama thanked them. It was during the time Governor Wallace had announced, "Segrahgashun now, segrahgashun tommorah, and segrahgashun *farevuh!*" So we were pretty lucky because we never knew if they would let us use their restrooms or not. I did a lot of driving, and I did a lot of praying along with the driving. That's what we had to go through.

Endnotes, Chattooga County, 1950-1958

1. JR Rosser received a Professional Principal's certificate (P-5), valid for life, in 1957, and a Provisional Principal's certificate (P-4) in 1951. Additionally, he had a Life Professional Elementary Teachers certificate issued in 1950. These credentials were issued from the State of Georgia, Department of Education. They were based on a four-year degree from an accredited college plus additional hours in specific areas, as well as a specified number of years of teaching and administrative experience.
2. The term *so many* meant a specified number in a formula or a minimum requirement.
3. Prince Hall Masons, a Black fraternal order.

4. Mr. Charles Aycock, principal of Rome High School, the city of Rome, in Floyd County, was a friend of JRR, Sr. He was also Audrey's piano teacher for four years.
5. It was easier and quicker to drive to the town of Summerville from across the mountain in Gore than to the community of Holland.
6. D. D. White worked at a school in Dallas, Paulding County, until his retirement (Personal conversation with Mrs. D. D. White, February 1, 2007).
7. At the beginning of each school year, principals registered the "class" of their high school.
8. Previously segregated activities where Black schools had different names for the same extracurricular activities were swallowed up by the White organizations. The Georgia Educators Association and subsidiary organizations controlled everything pertaining to high schools. The Black leadership structures in operation and historical records were destroyed because they were believed to have no value.
9. Four-lane highways divided by a median of shrubs and trees.

HABERSHAM COUNTY

Year	Age	Education	Events/Community Service	
1958	52	Audrey enters Ft. Valley	SY 1958-1959	
1959	53	Suejette earns BS degree	SY 1959-1960	Chamber of Commerce (CC)
1960	54		SY 1960-1961	Biracial Committee
1961	55		SY 1961-1962	
1962	56	Audrey earns BS degree	SY 1962-1963	
1963	57		SY 1963-1964	
1964	58		SY 1964-1965	
1965	59		SY 1965-1966	
1966	60	Assigned to Alto Institute	SY 1966-1967	Cornelia Planning Board
1967	61		SY 1967-1968	
1968	62		SY 1968-1969	Local Selective Service Board
1969	63	Retirement	SY 1969-1970	Member GMAPDC
1970	64			

Chapter 24

THE REGIONAL SCHOOL

Mr. Hallford, superintendent of schools in Habersham County, had just died, and his wife was carrying out his term. He and she were instrumental in building the regional school. They had changed directors at the office of Negro Education. The last director was in office, and she called him to say that she needed a principal for the Black school. He told her, "I know of one whom, if you can get him, he will do the job." He gave her my name.

She called me on that same phone we used for our program in air defense. When I was talking to her, the lightning hit the wire and knocked off the phone and followed the wire into Holland School. It terminated the conversation. Before the lightning struck, we had both made arrangements for me to go to Cornelia for an interview.

In August 1958 we moved from Lyerly to Cornelia. No White realtor would deal with me about a house. There was an old barracklike building[1] directly in front of the Baptist church. It was on what was Soque Street, which they named Martin Luther King, Jr., Drive. My wife and I had to live in that until something could be built.

All those years we were in the Holland community, we had to live in a converted army barrack that was on the school campus. We wanted a house with bedrooms by the time our daughters came home for Christmas. They were both going to school down at Fort Valley State College. It was Sue's senior year and Audrey was a freshman.

The only Black man that had land to build on was a trustee at the school. He built me that little cinder block house down there on Elrod where Bobby Pickins is living. He promised to sell it, but every time I wanted a price from him, he said he hadn't figured it up yet. I didn't want to be in anybody's debt, so I told him I'd pay him a hundred dollars a month until he got it figured up. Then we'd make some arrangements on paying the rest. After a few years or so, I decided to force his hand on a price.

That's when he told me he wasn't going to sell it, but he'd rent it to me, and he thought a hundred dollars a month was fair. So I could just keep on doing like I was doing. We didn't have a written contract. I trusted him, but he reneged on his word. I told him I had paid all these months because I thought we had a gentleman's agreement about buying the house. Since he wasn't selling it and had canceled our agreement, then he could consider the money I had already given him as rent paid in advance. So I didn't owe him anything for about the next two years.

I went on to Mr. Cliff Kimsey at the bank and bought this land. It was woods and a dirt road. I contracted with Paul Reeves at Habersham Hardware to build this house. I built me a brick house. When I moved here, there wasn't a Black man living in a brick house in this whole county. I had opposition among the Colored there. Some said that I thought I was White because I was in a brick house. They bet I wouldn't be able to pay for it. That was the rumor.

I was principal of the comprehensive school[2] for Blacks in Cornelia. They had a trustee board of six or eight Black men. Trustees furnished wood for the school. We didn't have much need for wood, so they were dismissed.

It was a "brick school" built for Blacks as an effort to avoid desegregation of the White high schools. Four counties had pooled their building funds several years before and built a school on the edge of the Black community in Cornelia. Banks, Habersham, White, and Rabun County boards[3] created this "regional school" in Cornelia.

When they were putting up the building, I guess they figured a gymnasium was too good for Blacks because they gave it to the White elementary school instead.

They hauled the Black high school children from these other counties to the regional school. Families with Black students of high school age from outer parts of Habersham County and Black high school students from the three other counties were bused to the school. Enrollment in the high school was therefore greater than that of the elementary school. Three families of students were bused from Rabun, one from Banks County and one from White County. In two cases there were busloads of students—one was a small bus.

When I presented my P-Five certificate to the superintendent, it just about "set the woods afire." Hadn't anybody seen one before. They didn't know what to do. Then everybody had to "light out" back to school. I didn't know at the time that it was having such an effect. I didn't know until later that I had stirred a hornet's nest.

The president of the board of education told them in a meeting of the AARP that we are paying Rosser more money than we are paying any of the other principals because of the supplement[4] for the number of teachers supervised. My rate of pay was higher than anyone in Cornelia, including the superintendent. They all went back to school. The superintendent had a sixth-year certificate, so she and all of the principals went back to school. This was somewhat significant.

It was about the same time that someone offered that "new math" approach. That was my first year teaching here. I took a course on the new math during the summer. It depended on what base you were working in. If it was base ten, then nothing much had changed from the old math. I never got much out of it. I couldn't use it. The way you checked to see if it was right was by the old method. So it soon dropped out. When I found out that you had to use the old math to check the new, I quit fooling with it as soon as I got it.

The school was segregated until the fall of 1966. After this plan to keep from integrating schools failed to meet federal requirements, counties were forced to integrate. Mrs. Nell Hallford was still the superintendent. She was so scared of this situation. She was a White woman, and she said it wouldn't work to let any of the Black teachers from the segregated system teach White children. She told me to tell my teachers that none of them would have jobs for the next school year. Oh my!

The desegregation plan called for converting the comprehensive regional school into a school that housed only primary grades[5] for the city of Cornelia, White and Black. So the building became an integrated school housing primary grades.

When schools were integrated, some Black people thought that I was selling them down the country by letting Whites take away the school when I was principal. They thought I had let the city take the Black school. They said I had sold my people "down the country," and like that. I had all of that to put up with. Nobody asked me the truth about the situation, and I just had to go on through with it.

Mama was the only one of the Black teachers who was accepted to continue in the new program. The White fellow who was her former principal, Broadus Quarles, interceded for her. He told them at one of the principals' meetings, "You don't know that it won't work unless you try it, so you ought to give her a chance." He gave her a chance. I would have "raised up" if I had known at the time that all of that was going on. But I didn't know it, so I didn't interfere with it.

The first year no Whites wanted their children in her room. There were four first grade teachers. So the parents felt like they ought to have a say-so about who was to be their child's teacher. They didn't. She was assigned children by the principal the same as the other teachers. They soon found out she was an excellent teacher. After that word got around, they all wanted their children in her class.

After Mrs. Rosser was teaching first grade for a while, a federal investigator visited the school and, while speaking to her, said, "I hear you have a Black teacher in this school, can you point her out to me?" She, of course, identified herself. The fellow was a bit chagrined. Mrs. Rosser did not look Black to either Blacks or Whites. This gives you some idea of how much they did not want "black" teachers teaching *white* children.

Mr. Quarles and I had become friends. He was having a problem with some of the Black boys talking back to the fifth-grade teacher. He asked me what he should do. I told him they needed a Black teacher. I advised him to try to get Mrs. Sarah Cook on his faculty.

I think she was working at a school over in White County, but she was living in Cornelia, and she might like to work at home again. She was one of the teachers that had taught fifth grade for me at the regional school. He did manage to get her over there. And he didn't have any more trouble with the little Black children misbehaving.

Chapter 25

RACIAL RESIDUE

It was along in 1960, or there about, when the folks were trying to integrate, especially these eating places. We heard of a group of Blacks coming this way from Atlanta. They got to Gainesville, twenty-four miles south of here, and stopped to get gas. While they were stopped, they asked to use the restroom, and they were denied use of it. That was routine. Folk could read about Blacks buying gas and being denied use of the restroom and café.

To deal with that situation, our mayor appointed a biracial committee. Blacks in the county represented less than 4 percent of the total population, out of about thirty thousand. I was quickly appointed to its committee.

I was the secretary, and a White minister, Reverend Marcus Martin, was the chairman. The committee functioned several years. It was funny that other Blacks and Whites were appointed to the committee, but they never did meet. Of all the years the biracial committee met, I never saw but two meet with the mayor, only the chairman and me.

Others were appointed but were scared to come, the Black ones, that is; while the Whites sort of ig'ed[6] it, I think. I knew a few Blacks who never showed up. Rev. Joe Banks was one. Washington Irvin Gober was asked, but he didn't want to have anything to do with it. I can't remember any of the Whites who didn't show up, so I don't know if I knew them or not.

We heard that they were coming out here from Atlanta. We discussed several situations and plans. You see, at that time, the Blacks could go around to the back of these eating places, and they'd serve them through a window if they wanted something to eat; but they couldn't be served inside. It was like the drive-through places now. Only Blacks were on that track; the Whites went inside to sit down. We thought that was bad then but would much rather do that now.

We got the idea of telling the folk about a plan. At that time there was the Steak House and a few others, but there was no McDonald's or the like. It was mainly the Steak House. Reverend Martin was a Methodist preacher. He was pastor at the First Methodist Church up there on the hill. It was not named United then, just Methodist. He was pretty liberal, and of course, he didn't object to Blacks eating in there. Reverend Martin and I went to see Bobby Joe Caudell. He's still the owner of the Steak House.

We just told him the alternatives. We said, "You might as well think about serving these people when they come 'cause that might avoid some trouble that could start if you don't." We said, "You can throw away the plates after they use them, if you want to, but serve them, just for peace's sake." We asked him to think about it.

Sure enough, a group of fellows from Atlanta did come. Two Whites and three Blacks came into the Steak House and asked to be served. The little waitress ran back in the back to the fellow that owned the Steak House and said, "It's some Black folk in here and they want to be served, what're we gonna do?"

He said, "Take their orders and serve 'em! Then give 'em the bill!" So they served them. Now, I don't know what they did with the plates afterwards, and I don't know whether on not they overcharged them on the bill, but they served them. Those fellows went on their way, and they didn't come back anymore.

There weren't but a few cafés around here—the main one was the Steak House. When it agreed to serve Blacks, the others followed suit or followed his lead. It was inevitable that they were going to have to serve Blacks for their businesses to keep up.

The biracial committee had warned the proprietors that cafés should expect such action, and they encouraged the restaurants to accommodate the nonlocal patrons. It was just like the superintendent of schools: Whites were afraid of integration. They didn't know what would happen. *Once upon a time*, like some of the restaurants, public meeting places were also White *only*. They are now open to all races.

The next year Reverend Martin went from here to Roswell, a place between here and Atlanta. He and I had become friends, and once I stopped by there to see him. He was the proudest thing in the world, with me coming from Cornelia to pay him a visit.

I saw him one other time. Once he came back here, and we got a chance to see each other. I don't remember what organization it was, but they were holding a convention here in Cornelia, and he was affiliated with the group.

A man named Bruce was appointed chairman, and I was still the secretary. I can't remember his first name. I had all of this written down in minutes somewhere down in the basement, among those papers that were destroyed when it flooded.

Mr. Bruce was sort of liberal minded, and we got along. He was the manager of this place up here, this plant. I don't know what they manufacture,[7] but it's been there all that time. Bart Scott, my neighbor, worked for him. As a matter of fact, Mr. Bruce built that little block house for Bart that he lives in down there.

The mayor that appointed the biracial committee was Herbert Richie. He was sort of a friend of mine. When I came here, he was in the milk business, furnishing milk to schools. When I took over as principal at the regional school, that school was about three hundred dollars in debt to him. I knew it was an honest debt, so I went to work and paid it off. He told me that was some money he thought he was never going to see. So he became a friend of mine.

After he was mayor, he ran for state senator and didn't get elected. He smoked a pipe all the time. When he was "laid out" in the funeral home, he had on a tie that had a burn hole in it. His wife told me, "Mr. Rosser, you see that hole. That's from his pipe." I believed it because he'd always hold that pipe in his mouth.

The next man, after Mr. Richie, was another what you might call *liberal* mayor. Along about the same period they integrated eating places, they had to integrate the schools. They did all they could to keep Blacks and Whites in separate schools. Sometime this worked to our advantage. No Whites lived between here and town, so there were no sidewalks here. The only sidewalks were in the White part of town.

The little children had to walk to school in the street. They would walk together, side by side, and they would have the street full. I had to go to the city government to get a sidewalk for the schoolchildren. I told them, "The street is so crooked and hilly and there is no sidewalk. If one of them gets killed, the city is going to be held responsible." They listened to me; that's how we got that sidewalk.

After the sidewalk was there, I had to teach the children to stay on it. I told them the city wouldn't be responsible for them getting killed by a car or run over unless they stayed on the sidewalk. They pretty much stayed over there after that. It is still the only paved sidewalk in the Black community. It starts at the point of Third Street and Elrod and goes down the hill and up the hill to the schoolhouse and stops. I seem to remember hearing recently about them paving or going to pave a half of block of sidewalk from the park to Third Street.

They never did call me that, but I was like the "chief public relations officer" for the Black community. As I was principal of the Black school when the school was taken over by Whites with desegregation, I had that distinction and had contact with the city council and chamber of commerce. Other than somebody meddling, I was the only Black person who had any connection with the White power structure for about twenty years.

I made some efforts toward a public meetinghouse in the Black community. My objective was that Blacks needed to learn how to get along with the other group. Some Blacks, like Howard Moss, thought people who were advocating interracial harmony were selling them down the country. After the integrating of the school, I had told the Cornelia City Council that, with the closing of the Black school, the community was left without a public meeting place. And the community was falling apart. Public nonsectarian programs could not be held.

The city proposed to build a Black community house and appropriated money for it. Then the cost was eighteen thousand dollars to build a brick building with an auditorium in it. The city agreed to build the building if the Black community would put in the chairs.

The newspaper had printed plans for a new building with a gymnasium and heated pool. It was talking about planning to build a pool that could be heated for year-round use. Blacks heard about government money for a heated pool and park. This rumor came before the money was appropriated for the community building.

A Black fellow, who was a deacon in the Baptist church here, was opposed to building a "community house" in the Black community, for which the city had appropriated money. The deacon thought that the Cornelia gymnasium project was going to be a federal-government-sponsored project. He thought that the city wanted to put up a little shack in the Black community so we would be barred from the gym. He felt the building that the city proposed was designed to "torpedo" this one funded by federal funds.

The mayor also appointed me to the Municipal Planning Board. The Municipal Planning Board was having a meeting. The deacon and another Black man came to the meeting and expressed their opposition to the "Black community house." They had gotten other Black people in the community to sign a petition against it. After the deacon rambled around, the board asked him what did he want. He said, "We want to wait on the federal money."

He was thinking that if they put up a small building in the Black community, they could deny Blacks participating in the swimming pool. He thought he was smart, but he thought like a fool. He had torpedoed the community house. Neither project was built. There was quite a commotion for a long time in the Black community following that action.[8]

To this date, March 2001, that "proposed building" has not been built. A few four-family units of cheap rental housing were built in the Black community with federal funding. There had been a little swimming pool across from Shady Grove. The city filled in the pool; put up a little shelter, swings, and a sandbox; and called it a park. It was a recreation park for Black women who took care of White children and would bring them there.

In 1968, about the time Martin Luther King, Jr., was assassinated, the mayor appointed me to represent the biracial committee with a group of White ministers. I had retired from most of the activities, but I introduced the first Black minister to the group. He was Reverend Terry McCreary,[9] who came along later as a member of the Ministerial Alliance.

For two or three years he delivered the radio message every morning for a week. He was the only Black minister who joined the group who took turns preaching for a week at a time in the mornings over the radio.[10] Before that time, Whites did not know that Blacks could do anything like that.

PETITION

STATE OF GEORGIA /
COUNTY OF HABERSHAM /
CITY OF CORNELIA /

DATE: 5 May 1968

On Thursday Night, May 2, 1968, at 7:30 P.M., the Citizens of the above city met to discuss the proposed plans for construction of a "Community Center". After a lengthy discussion of the aforementioned; We, the undersigned, without prejudice of any sort, concluded to draw up this PETITION against the proposed plans. Our conclusions are based on the following items:

FIRST: SIZE AND LOCATION OF PROPOSED BUILDING NOT SUITABLE.

SECOND: DISTURBANCE TO THE NEIGHBORS.

THIRD: PROPOSED SITE FOR BUILDING NOT TO BE CHANGED FROM RESIDENTIAL TO COMMERICAL AREA.

FOURTH: WE, THE CITIZENS, WILL WAIT ON FEDERAL FUNDS.

[signatures]

Petition, City of Cornelia, Habersham County, Georgia

Chapter 26

RETIREMENT

After the comprehensive school closed due to integration, I had an option of working two more years to qualify for a pension. The Alto Institute was a reform school for boys, but it had a high school operated under the Board of Education of Habersham County. That was the reason why the board sent me down there. I was not principal at the reform school. There was a White principal. I was a teacher of basic mathematics, algebra, and geometry.

When I started teaching down there, I knew that the inmates were dangerous because they weren't sent to prison for singing too loud in Sunday school. So I kept them in front of me. In fact, I fixed my seat. I put my seat and desk in the corner where nobody could get behind me and I could see all of them in the room at the same time.

Of course, I didn't do much sitting down. To the extent that some of them really questioned me about it, I had to tell them that I didn't have any eyes behind me, so I had to be able to see them because I knew what they would be able to do if they got a chance. They really did jump the principal and beat him up. I didn't know what was going on at the time, but two or three of the inmates went into the office and got behind the principal and sort of roughed him up a bit. This group was a mixed-race group that had developed an attitude toward the situation.

I often reminded the inmates that they were sent there as a consequence for something they had done, and they had to obey the rules without question. In the other instance I'd let the students discuss things. That's the only way to get something across, to talk about it. They'd always bring up the negative side, and I'd give them some pointers to think on.

I had one fellow in there for stealing cars. I just took it apart and asked, "Why did you steal cars?" He said, "Just for the heck of it." I told him he needed to get some new "hecks." That was my judgment. Maybe he accepted it, maybe not.

There were "Black" classes and "White" classes before the state made us integrate the prison. In other words, White teachers teaching Whites and Black teachers teaching Black boys. We had a "White team" and a "Black team" that didn't have anything to do with each other. The second year I was there, the state made us integrate.

My task, at least I took it to be my task, was to promote the integration. And of course, I told them the situation. I told them when the White boys came in there, "Now, we don't understand each other so well. We've got to do what the *state* tells us to do. I've got to learn that when you say *tar* that you are talking about an automobile *tire*. You have to know when I call it *tire*, you call it *tar*, that we are talking about the same thing."

I had several situations like that that I would use. I told them that we have got to understand. "I want you to understand what I say up here on the board. If I tell you something about a *cat*, and if you don't understand the word c-a-t, I will draw one on the board for you." That was the way I went about teaching.

At the beginning of the day, I had group counseling, like in the first five to ten minutes of the day.[11] I would tell the students, "I know you all will be glad when you can walk out from here a free man." I had some rejections of that.

One person told me, "Mr. Rosser, you just don't know. Ever since I have been here, I have gotten three meals a day. It wasn't always what I wanted or would have liked to have, but it will keep you alive. Before I came here, sometime days went by and I didn't get anything to eat. I don't know whether I want to be free or not."

The idea was to rehabilitate the young men who were in the reform school. In fact, it was a life situation down there. During the counseling sessions, I told them about being honest and doing the right thing. One fellow told me, "Mr. Rosser, I want to tell you something. When I came up here, the only way I knew how to unlock a lock was with a key. *Now,* I can *pick any lock* on the campus." That showed me that there wasn't any rehabilitation going on. They were learning "devilment" from each other and promoting their devilment and the like.

There were some inmates who wrote me after they left the reform school. I had a situation almost like Mama's was here. The fellows really preferred me. Sometime the principal had to get in behind them to drive them out of the restrooms into some of the White professors' classes.

In fact, when they jumped the principal, it was for chasing some of them out of the restroom who were avoiding going to a White teacher's class. Some inmates wrote the superintendent and made some complaints about the principal there.

There were two or three of us ex-principals who were teaching. One of the White teachers who taught along with me wrote the superintendent and told him that he didn't like what the principal was doing.

The way I call it is, we were all rejects from the public schools, teachers, and students. Two more of the White principals that had been rejected from the "White" integrated schools were down there at Alto. So in fact, we were just a "bunch of rejects," all of us that were teaching there. But the students had their preferences.

Some inmates wanted to get out and would do anything to get out. They did not express to me they felt they were wrongfully convicted or anything like that. It just was not to their liking. If they could escape, they would do that. There were many escapes.

There were one or two fellows who did not want to get out. Sort of like the one who was talking about getting free meals. When their time was about to expire, they would do something to extend their time. Two fellows climbed up the water standpipe[12] and refused to come down. They had to send for one's mother to talk him down. He finally came down and was sentenced to a longer term. Both of them were.

I was at Alto three years and almost two months. The year I became sixty-two, I put in for retirement from the Georgia Teachers Retirement System. I did not immediately stop teaching at the reform school until my state pension started to come in.

The state attorney general ruled that I would have to pay back the two months' salary that I had been paid after my pension began. The cost of living has now increased to the extent that my retirement pension is more than doubled what it was then.

Endnotes, Habersham County, 1958-1970

1. The building had served as the elementary school for Blacks in Cornelia. It was selected for housing because it had a "bathroom." The only other house available nearby did not have an indoor facility.
2. This was a combination of elementary and high school This was a common design of schools for Black children in small towns: grades 1 through 12 in the same building, on different wings, sharing the same lunchroom and auditorium.
3. The County Boards of Education.
4. A principal's salary was, first, the base pay for a teacher computed on the individual level of certification and the number of years of experience, the same as other teachers. This base pay plus an amount set by the State Department of Education, times the number of teachers in the school, determined the salary.
5. Grades 1 through 3 were in Cornelia Elementary School #2 (Black school renamed). Grades 4 through 6 were in Cornelia Elementary School #1, where the gym was located.
6. Ig'ed is a slang or *ebonic* expression meaning "ignored."
7. They manufactured T-shirts. The plant is now closed.
8. According to Suejette, "Dad sold his two lots to Shady Grove Baptist Church, on which they built a parsonage and a trailer park."
9. Pastor of Shady Grove Baptist Church.
10. WCON 99.3 FM.
11. Modeled after the homeroom period in high school.
12. Another name for a city water tower.

GOLDEN YEARS

Year	Age	Education	Events/Community Service
1970	64	SY 1970-1971	
1971	65	SY 1971-1972	
1972	66	SY 1972-1973	AAA Advisory Council
			Chamber of Commerce Board of Directors
1973	67	SY 1973-1974	Habersham Retired Educators Assoc. chartered
1974	68	SY 1974-1975	
1975	69	SY 1975-1976	AARP Chapter #2040 incorporated
1976	70		Pauline retires
1977	71		Jury commissioner, Mountain Judicial Circuit
			Young-at-Heart Club
1978	72		
1979	73		
1980	74		GMAPDC Board of Directors
1981	75		
1982	76		HREA President
1983	77		
1984	78		The Bahamas
1985	79		Cancer Society Fund Chairman
			Hawaii, Golden Apple Club
1986	80		
1987	81		
1988	82		Honorary doctor of letters
1989	83		
1990	84		J. R. Rosser Scholarship founded for future teachers
1991	85		
1992	86		
1993	87		Ends last term, Chamber Board of Directors
			Ends last term, GMAPD Board of Directors
			Ends last term, AAA Advisory Council
1994	88		
1995	89		
1996	90		Ninetieth birthday celebration
1997	91		Emeritus member, HREA
			Stops driving

1998	92	
1999	93	Stops teaching Sunday school
2000	94	Pauline dies December 15
2001	95	Attends last meeting, AARP
2002	96	Attends last family reunion
2003	97	Attends December meeting, HREA
2004		Died May 26

Chapter 27

CHAMBER OF COMMERCE

I was the first and only Black member of the Habersham County Chamber of Commerce for quite some time. Because I had to buy a lot for that lunch program, I joined the chamber and paid my fee of fifty dollars.

Let me tell you how I got on the board of directors. At the time the chamber was having a little difficulty with the county commissioners. The county commissioners appointed me to the board of directors of the chamber of commerce.[1]

I stayed there until after I retired. The commissioners would not release me from the board of directors at the time I reached the age of retirement, so I just stayed there as chaplain of the chamber, some twenty-one years, longer than anybody else before or since.

Practically every business in the county was a member of the chamber of commerce. They paid dues as a fee of about fifty[2] dollars per employee in their organization. The custodial workers and cafeteria workers in the school were considered employees within the fee structure. Thus the permit to build the regional school had required that fees be paid for the nonteaching personnel.

We had low unemployment levels. The purpose of bringing businesses to the county was so that our citizens could find employment. Some of the products manufactured in Cornelia were women's clothes made by Habersham Mills, a garment factory; and, Lumite or something like that, they made steel storm doors and windows. I remember one[3] that made Johnson & Johnson health care products. We gave them two or three years free of taxes to get established.

County commissioners are elected on the same years as the presidential election. There are five commissioners. They regulate commerce and give the companies free port[4] on their products for a certain period of years, to come into this county. They could manufacture their goods and not have to pay taxes on them for a certain length of time.

I was appointed to the Public Service Awards Committee along with three women and one man. The committee would have to get certain information from the persons who received the award without formally seeking such.

The committee did its own research and did not reveal the recipients until the right time before the annual banquet. One of the ladies, Elizabeth Kimsey, was

135

the leader who could get information from a person without him knowing he was giving out anything. We told her jokingly that she could pick a fellow's pocket and he wouldn't know it.

Our first award was to the president of Habersham Mills because they had hired Colored people and given so many jobs. They were almost like the railroad was in the Tallapoosa area. I was involved in the selection process. There were several awards after that.

Each year I invited Richard and Sammie to be our guests. The banquet was fifteen dollars per person. I would pay for Mama and me and our guests. The members of the chamber knew Richard by his yearly visits. He was then the director of Emergency Management for the City of Athens and Clarke County.

I was a member of the Public Relations Committee when Colonel Sanders came to Cornelia to teach people how to make gravy for the new KFC restaurant in Cornelia. He was surprised that his group of employees did not know how to make gravy before he showed them.

I was the only Black person who was invited to sit down with him at dinner. Several bankers and other businessmen were also invited. They were all much younger than I was. I was about the age of Colonel Sanders. They asked me what I attributed my age to. As old as I was, they did not know anything about a Black man with a college education until I came to Cornelia.

Chapter 28

BEING A GOOD CITIZEN

I always tried to be a good citizen. As I got older, I figured God let me stay here this long for some reason. So I made donations to the Red Cross when there was a disaster. I joined organizations in the county here, such as the Retired Teachers. I served on the grand jury when summoned and was a part of several programs in the adjoining counties. I was with some longer than others but tried to do my part in assisting the community[5] where I live.

Just after I retired, we had a food program that was government funded. I was one of the drivers. My car carried the people from Cornelia up to the site in Clarkesville[6] for the meal. The food coordinator, Art Smith, had trouble with Norma Allen. She just took over, and he told me since she wanted it so badly, he was leaving.

I told him, "Maybe you could organize a chapter of the AARP." He wanted to know more about it. He and his wife talked with the banker, H. M. Stewart, and his wife, Helen Stewart, over the phone. That led them to organize the AARP chapter #2040.

W. W. McCray was the first treasurer. He served four years; then they elected me. I did it for four years. We got a caterer to cook for us to meet and eat. I was elected president. The man who succeeded me as treasurer lived on Black Snake Road. He was later elected president of the chapter.

In 1985 I was appointed chairman of the Habersham County Cancer Society. The purpose was to raise money. I suggested the "One Hundred Society" be formed. The judge's son thought that was too much and might embarrass some if they couldn't pay it. I replied, "I'm the only Black here, and I can afford a hundred dollars. There is no use being White if you can't pay a hundred dollars." I spoke before I thought, but that didn't make anybody angry and didn't make everybody pay it.

Mr. Voorhees Black and Mr. Whitfield helped me to raise money for that year. Pauline and I opened a savings account at Habersham Bank because Mr. Black had helped me with the cancer society.

14, 1985, Page 2

Receives Award

Rea Biggert, assistant vice-president for Field Services of the Georgia Division of the Cancer Society, presented an award to J. R. Rosser, for his outstanding leadership, service and faithful contributions to the Cancer Society. He was presented this award at the Annual Chamber of Commerce Membership Banquet held last Thursday, November 7.

Article from the TriCounty Advertiser, Weekly Newspaper

I was affiliated with the Georgia Mountain Area Planning and Development Commission [7] for more than twenty years. Thirteen counties comprised the GMAPDC. It was one of ten such commissions in the state. Two persons were chosen by their county councils to serve two-year terms on the commission.

Initially, one of the two persons was to serve a one-year term. I was selected to serve a two-year term as a member of the first commission.[8] The commission of twenty-six members organized by electing a president and vice president annually. For most of the time, I was the only Black[9] member of the commission.

Meetings were held monthly, alternately at various locations within the several counties. When we met, we would meet one day and complete our business. The annual conventions met two or three days. The main office of the commission is in Gainesville. Dr. Sam Dayton was the executive director.

The purpose of the commission is to recommend or oversee job training programs and opportunities for young folk, and adults too. It's supported by the Job Training Partnership Act[10] of the US Department of Labor. Education and training for youths and adults, including the elderly, were the focus.

We had programs such as meals on wheels for the elderly. We looked out for illiterate youths, provided education and job training and employment. Blacks in this area still comprise about 4 percent of the population.

I was a little bit ahead of the field by being an educator. And I was popular. I remember once this one fellow moved away, they decided to replace him—someone nominated me.[11] In the nomination process, I told them I had two choices. One was to "live fast, love hard, die young, and leave a beautiful memory."

The other was to "live so long, your head gets bald, and you get out of the notion of dying at all." And I chose the latter, so I would do the best I could. I didn't feel up to it, but would accept it if they would agree to my terms. They made some complimentary remarks, so they must have accepted my terms. I was voted in. Those things are most meaningful.

One or two times we had elections of officers, and there was competition. We got along pretty well, but the friction was coming. It would not have gone like it did if it had not been for my effort. I had the responsibility of having the number of votes to get a man elected. I guess that was a major accomplishment during my tenure.

Others were slow about expressing accomplishments. They were not city people, but country people for the most part. Everybody seemed to be accommodating. Dr. Dayton would ask for something, and some member would make a motion to accomplish that action. Some member of the commission would jump up and say, "You got it."

He operated from an agenda. I sometimes wondered whether "his way" for spending the money was the way that he should have spent it. There were state audits, like they did for all federal programs.

I left because my eyes started going bad. All our meetings were at night, and I couldn't see to drive. I was able to attend a little longer. Dale Bryant and his wife, Betty, would take Mama and me, in my car, for several years. When his time expired and he didn't have to go to the meetings any longer, I had to drop out too.

I've been away from it about eight years now. I think they made some changes, even changed the name.[12] Pat Viles was the director. His secretary, Pat Riley, a quite respectful young White lady, was there when I got there and was there when I left.

Endnotes, Golden Years, 1970-2004

1. According to Ed Nichols, executive director, JR was a member of the chamber from January 1964 to December 1985. There were no records available prior to 1960. According to a letter in his personal effects, JR was a member of the board of directors from January 1972 to December 1993. The chamber and the Habersham County Board of Education cosponsored an award in his honor in 2005. The JR Rosser Award continues to be given annually to the outstanding educator in Habersham County and is now solely sponsored by the board of education.
2. Amount in formula could not be verified due to unavailability of records.
3. Ethicon, a medical corporation, is a Johnson and Johnson Company.
4. Free tax.
5. "Mr. Rosser was on the Planning Board from 1966 to 1973, then he served another time after that. He was Elections Clerk, Manager, then Chief Manager between 1971 and 1977 and was on the Board of Adjustment from 1976 to 1985. He did a lot for city government. I wish I could've known him" (Personal communication, Jeff Barrow, Director, Cornelia Planning Department, March 2009).
6. Mrs. Jean Mabry followed JR Rosser as chapter president. She remembers being a driver in a Senior Citizens' food program that had a site in Cornelia. She says, "Dr. and Mrs. Rosser were both very fine people. I remember he would call me his 'Little President.' He was always so encouraging. He had been president, and he knew what I had to do" (Personal communication, January 2009).
7. The GMAPDC was formed in 1962 by the Georgia Department of Labor.
8. JR Rosser served on the board of directors 1980-1993. He was on the Executive Committee for several years as chaplain (1983-1993). He represented Habersham County. The other twelve counties were Banks, Dawson, Forsyth, Franklin, Hall, Hart, Lumpkin, Rabun, Stephens, Towns, Union, and White (Personal communication, Peggy Lovell, administrative assistant, GMRDC, January 2009).
9. Mr. LJ Harrison says, "Mr. Rosser was the second Chaplain elected to the GMRDC Board of Directors, and I am the third. I've been there since 1990." Mr. Harrison is a Black representative from Stephens County (Personal communication, January 2009).
10. "Dr. Rosser was one of the original members of the Georgia Mountains Private Industry Council. This group gives direct oversight to JTPA, and is accountable to GMRDC, which was originally the GMAPDC. The Area Agency on Aging is also accountable to the GMRDC and gives oversight to PEAK Services. I believe he served on all of these boards, at one time or another. Back in the 1980s he advised me to spend money on computers,

because that was going to be the key to progress. He believed that education was of the utmost importance in these JT programs" (Personal communication, Harvey Clanton, first Black director of the Job Training Partnership division in the GA Mountains Region, January 2009).

11. This was apparently the Advisory Council for the Division of Aging Services known as the Area Agency on Aging (AAA). He served as chairman/president from 1982 to 1993 (Verified by Peggy Lovell in consultation with Pat Riley Freeman, January 2009).
12. In 1989 the Georgia legislature passed an act changing the name to Georgia Mountains Regional Planning Development Centers (GMRPDC). This did not change the representation from the same thirteen counties but expanded the number of commissions in Georgia. In July 2009 the name will change to the GA Mountains Regional Commission. It will be one of twelve commissions, and the same thirteen counties will still be represented.

PART TWO

Reflections

Chapter 29

MAMA'S GUIDANCE

June 29, 2001
Pauline Lynch Rosser was born June 29, 1913 and lived until December 15, 2000. This day would have been her eighty-eighth birthday.

JRR: I was just thinking about things Mama did to help me. Mama was a very intelligent person. I observed that early in our lives. And of course, I didn't make any big decisions without her okay.

ARM: What was the first time you can remember Mama making a difference in what you did?

JRR: I had problems with my education. When I got two years of college, I was ready to quit my education. I had three boys. Mama told me I ought to go on with my education. She said, "When those boys grow up and go to college, folk might inquire, 'Which college did your dad go to?' And what would you say?"

A: So you did keep going.

JRR: She encouraged me to go on. So I scuffled along for a few years more, and I got my college degree. I didn't have any time off. I was doing correspondence work and both sessions during the summer, and all like that.

I had difficulty doing my work, to tell you the truth. I had a course in child development, but they[1] said the school I had it from wasn't accredited at the time I took it. So they made me take another one, "child care."

I had to try to work, raise a family, and go to school. It was almost too much on me. Sometime I wished I could do like some other folk and go off into amnesia until it was over. Then come back to my right mind. Of course, I just thought it. I never did it.

A: Did you ever do anything that would make people know that you were fed up with it?

JRR: When they did let me graduate, I had planned what I was going to do at the ceremony. As soon as they awarded my degree, I was going to take off my robe and leave it right there. Then get in my car, leave, and not go back to Fort Valley anymore. But Mama and you children were in the audience, and

	I couldn't do it. It might have embarrassed you. Then Mama went there, and Billy² went, and Sue, then you, so I had to keep going back.
A:	You must have been tired of school, yet you went on to graduate school.
JRR:	Yes, I went to summer school that next summer. I didn't intend to go, but Mama encouraged me to go on to Atlanta University. My first year there was 1951. Then, of course, I went back the next two or three summers.
A:	Mama encouraged you to *keep going* to school. Do you remember some things she *stopped* you from doing?
JRR:	Mama stopped me from doing a lot of foolish things.³
A:	What were the most important situations where she made a difference?
JRR:	Our son, Richard, was kicked out of school in his senior high school year because he helped his other classmates with their lessons.
A:	How did he help them? What do you remember about that?
JRR:	Well, Mr. John T. Smith, that principal down at Moten High School, said Richard was giving them the answers to questions. I don't know exactly why. After he suspended Richard, I was ready to write him a terrible letter. But Mama advised me not to. And I didn't. I did go out there and talk with him and he let Richard back in school.
A:	Daddy, where were you living when the incident of the slaughter pen bridge occurred?
JRR:	I was living on Connecticut Avenue. All of our children had been born. You and Suejette were both born in that house.

When those White fellows were drinking and asked me for a match, I told them I didn't have one, and they told me to make them a match. I left and went up to the house and got my gun and was going to go back there.

They used a lot of profane language talking to me, and that is what made me mad. I never said anything about that part of it because Mama always advised me along those lines, times like that when I was doing something new.

When I think about it, Mama asked me where I was going, and I told her I was going to go back down there and shoot those fellows. She said, "Now you are going down there to tend to business with a gun and leaving common sense here with us, to be destroyed by what you do?" Because we knew that if I had shot any of them, they would have mobbed me, *and my family*. This has never been written down anywhere. |
| A: | How did it get named the slaughter pen bridge? |
| JRR: | The little creek that went down there had several streets that crossed it. That bridge was where the butchers would buy and bring a calf there, and they had these racks to hang them up to skin them and clean them off. And they had that business there. We called it the slaughter pen.

That bridge was across the creek, beside the slaughter pen. So they called it the slaughter pen bridge. They used the water to wash the animals. They would throw the intestines in the water for the fish, terrapins, and turtles to eat. |

It was just dusk when I was on my way home because I had borrowed Charlie Byrd's mule to plow up my potatoes. I had carried the mule back to his house. It was a little after sundown, and I was coming back home. I don't remember seeing them when I went in there to take the mule home. It was about a quarter of a mile down the little road down by the slaughter pen bridge, and Charlie Byrd lived up that road.

One of the fellows said, "Hey, nigger, give me a match. I want to smoke a cigarette." I told him, "I don't have one." He told me to make one. I couldn't ignore them because there was only one me, out there. I knew he was talking to me even if he did say nigger.

Ole man Brock's nephew was among that bunch. Dr. Brock was on the board of education. He told him, "That's our got damn nigger professor, let him alone." That is what made me mad, when the fellow was going to make me make him a match. They had guns, although I did not see such. And I was not going to face them anymore until I had mine.

This is why I knew they had guns. This fellow, Brock, and another buddy had been out all night, pretty soon after this occurred, and went home about daybreak, got into an argument, and both shot each other and died. This happened right there on Dr. Brock's porch.

A: What made you the angriest, what they said or the way they said it?

JRR: *Both.* Brock said, "That damn nigger professor." Profane language. That made me mad!

A: I knew that you used the slaughter pen bridge example when you told us to "think before you act." A number of times I have heard you say *that*, in a high school class or a Sunday school class.

JRR: Mama's the only reason I lived to tell about this. She was valuable to me, to our children, to others in the family. In fact, she was a credit to everybody who *ever* knew her and worked with her. She was a jewel.

A: Do you credit Mama with your living past the age of ninety?

JRR: This is what I've been saying! You see, my father died at age eighty-five. My grandfather died at eighty-five, and my uncle died at eighty-five. When I got to be eighty-five, I thought that would be curtains for me.

Suejette was having trouble stretching her money with her three daughters in school, and she couldn't give me the kind of present she wanted to, so she told me she would give me a party when I got to be ninety. I said, "What you talking 'bout, Sue?" I was worried about being eighty-five. But I did live to get ninety, and y'all gave me a party!

A: I know you had just seen some family members at the reunion in July. But even more of them came for your birthday. Were you surprised to see so many of your former students from Tallapoosa and Lyerly and Cornelia?

JRR: Oh my, yes! I saw some folk that I didn't figure I'd ever see again. It was a cold, damp night, but they came anyway. I was glad that so many came. I appreciated that. I have the tape, yet. Of course, Mama was there, and all five of y'all and our grandchildren, and I think there was a great-grand or two there, wasn't it?

A: Yes, it was. Other than seeing all the people, what did you like most about the party?

JRR: The president of Morris Brown College, Dr. Samuel Jolley, giving me a plaque. That was because I had given them ten thousand dollars. He said he recognized my name because he was in your class down at Fort Valley State College.

I really have been getting a lot of mail from them since a woman became president of Morris Brown. I am still worried by her sending me information whenever they are having a rally. She thinks that I was a Morris Brown College graduate. She doesn't know that I just finished from the academy.

I don't have another ten thousand to give them. I can't run with the "bigwigs" from other states and cities. I sent the donation because I appreciated what they did for me, taking me in and helping me finish high school, after ole man Hubbard kicked me out of Forsyth.

Chapter 30

TWENTIETH-CENTURY MILESTONES

May 27, 2001
Dad's responses were unrehearsed, unedited, and recorded. He had options to give a short answer or none at all. The events are not direct quotes; however those indicated by italics and page numbers in parentheses are from the Chronicle of the 20th Century, *Chronicle Publications, 105 South Bedford Road, Mount Kisco, New York 10549, (1987).*

Robert: Dad, this book is a chronicle of events in the twentieth century. I picked out a few that you might remember. I'll read the event, and you say whatever you remember about it.

JRR: Okay.

R: *December 26, 1908, Jack Johnson wins title from Tommy Burns by TKO, first Negro to win the world heavyweight boxing championship.* (119)

JRR: I remember that they had a contract out for him if he won that last fight he had. When the referee was counting him out, he looked over at the newsman standing near and laughed because they had said that if he won, they were going to kill him. He was looking as to say that "I could have done better." I remember the photograph.

R: *February 17, 1909, Geronimo, famous Apache leader, dies at St Louis World's Fair.* (121)

JRR: I heard about him, but I don't remember anything special.

R: *May 1, 1910, NAACP organized from the National Negro Committee.* (137)

JRR: I remember it. If there is any such thing as being on the fence, it was my opinion back then that it was too dangerous. I was oriented that "I was a little nigger boy." When the president was addressing a class and said, "You can be president of the United States," I didn't think he meant me. I thought it was because he went to the school and was a student there. I didn't have any such expectations.

R: *April 15, 1912, iceberg sinks Titanic, drowning 1,595.* (160)

JRR: I remember back there when the *Titanic* went down. I remember when the postman brought the paper. The postman traveled by horse and buggy. Some few people had mailboxes. If you didn't have a mailbox, the mail was left on

149

the ground by a tree. We didn't have a mailbox. He got out of the buggy and laid it at the root of a tree. He got to talking with my daddy about it. He laid the mail down because my dad always took the *Atlanta Constitution* paper. They had RFD, rural free delivery of the mail.

R: *Satchel Paige.*

JRR: I had "heard" of Satchel Paige, but I did not know him.

R: *March 10, 1913, Harriet Tubman died in NY at age 92. She was born on a plantation in 1849. During the Civil War she had been a cook, nurse, scout and spy with the Union Army in South Carolina. (170)*

JRR: Along in those days, unless Blacks committed a crime, we didn't make the news.

R: It is still almost that way [*laughter*].

JRR: Then you know what I'm talking about.

R: *March 3, 1915, "Birth of a Nation" shown in New York. Three-hour motion picture account of the Civil War and Reconstruction. All Negroes in the film are foolish, evil or both. Story by DW Griffith of a white Southern family searching for dignity within turmoil. Negroes and white liberals plan to protest the film. Ticket price was $2.00. (193)*

JRR: I heard of *Birth of a Nation*, but I did not see it. While I was at Forsyth, they learned how to have a "talking" motion picture. In the theater they would print over the top of the picture what the characters were saying. After that, they invented the "talking picture" along when I was a young man.

R: *August 31, 1916, $250 Fords rolled off assembly lines. (209)*

JRR: I noticed many Blacks having cars. The first I noticed who bought one was a Black doctor. Jordan was his name. He gained his fame because there was a rich White man's wife who got sick and the White doctors in Newnan couldn't do her any good.

He called in the Black doctor, who told him, "I can cure her." The White man gave him the okay to try. He treated her and she got well. So the White folk didn't allow him to treat Black people.

One night he had to go somewhere, and he wanted to see how much gas he had. The tank was under the front seat of the model T. He screwed the top off the tank and struck a match to see how much gas was in there. The thing blew up and killed him. It burned him up.

They fixed up that Ford of Dr. Jordan's. Another fellow, named Paige Mathis, at Lone Oak bought his old burnt-up car.

R: *May 18, 1917, President Wilson signed into law tonight a bill requiring all American men between the ages of 21 and 30 to register for possible service in the United States armed forces. Selective Service Draft Act. (218)*

JRR: I remember World War I. It ended about 1918. I remember the registration of men ages twenty-one to thirty-one. My Dad was over thirty-one when they began the registration.

R: *July 19, 1921, eight million American women hold jobs. (281)*

JRR: I remember a whole lot of discussions about women were going to vote and had a position out that they were going to rule the world. They were going to take over from the men. That was rumored back there when I was a young man. In biblical times when Esau and Jacob were born, their mother fixed some hair to grow on Jacob's arm to make him be like Esau. That was women taking over then.

R: *March 18, 1922, Mahatma Gandhi imprisoned for civil disobedience.* (289)

JRR: I remember him about the time of that other fellow, Marcus Garvey, who advocated going back to Africa. I used to hear a whole lot of discussions, but I didn't know how big the world was. They were talking about Blacks and Whites. Blacks weren't called niggers then. What Blacks said or did in those days, unless they committed a horrible crime, did not get front page in the newsprint.

R: *August 2, 1922, Alexander Graham Bell, inventor of the telephone, died early today at his home in Nova Scotia. He was 76. Bell was born in Edinburgh in 1847 and came to Canada in 1870, working as a teacher of the deaf. He began working on the telephone in 1875 in Boston and achieved success on March 10, 1876, when his assistant sitting in another room heard Bell's voice say, "Mr. Watson, come here. I need you."*

JRR: I remember that.

R: *February 1923, a young Negro woman named Bessie Smith has come out of nowhere to capture the imagination of people who like that kind of jazz called "the blues;" "Tain't Nobody's Bizzness if I Do" and "Down-Hearted Blues." She was a Native of Chattanooga, TN.* (300)

JRR: I remember Bessie Smith and Mamie too. She had a sister named Mamie. I heard all of their records. I was just about grown when I heard a graphaphone. I could sing those songs like "Casey Jones," "John Henry," "Uncle Bud" and "Jailhouse Blues." I can sing some of those yet.

I admired a few of those singers. Mamie Smith, Lemon Jefferson; they made records. I could sing like they sang. Hudson and I sang together at Forsyth and people used to like to hear us. We learned the songs on the graphaphone. When a new record would come out the man in the store would get it.

R: *November 1925, the flapper dress is the in thing.* (330)

JRR: Black women followed the styles just as White women did. When the "big apple" hat and the "zoot suits" came out, Blacks were wearing them. They wore them to church. Ralph Belcher and his brother Roland wore them. Whites did not wear them much.

R: *January 1926, a new machine capable of the wireless transmission of moving pictures was demonstrated to members of the Royal Institution in London today by John L. Baird, a Scottish inventor. Baird calls his invention "television."* (332)

JRR: We heard about it. But I remember when the radio first came to Atlanta. WSB. Welcome South Brothers. I remember when it opened up.

R: *November 23, 1927, President Coolidge has commuted the prison sentence of Marcus Garvey, the self-styled "Provisional President of Africa." Garvey, a controversial Negro colonizer has been imprisoned in Atlanta Penitentiary since 1925 after his conviction on a charge of using the mails to defraud. He will be deported to Jamaica as an undesirable alien. "back to Africa movement." (352)*

JRR: We had discussions on Garvey. He advocated going back to Africa. Some of our discussions were on the fact that we had never been to Africa.

R: *Duke Ellington and his band in Harlem. (354)*

JRR: As I remember it, that was the Black section of New York.

R: *October 24, 1929, Black Thursday: Stock market crash. (375)*

JRR: It didn't have much impact on me. I didn't know anything about the stock market at that time. But I remember talking about it. They didn't have television broadcasts then. The only thing that we would get was information in some magazines and the newspaper. The *Atlanta Constitution* carried a lot of news, but it did not cater too much to the Black program.

R: *March 31, 1931, Alabama charges Negro youths with rape. They have come to be called the Scottsboro Boys, nine Negro youths, all teenagers but one. They were arrested in Scottsboro, Alabama, on the complaint of some young white men, who said that the Negroes had driven them off a freight train in Chattanooga, TN, six days ago, all were drifters. (392)*

JRR: Yes, yes, that was everywhere. I have lived in a multicultural society all of my life, and I had problems in the late 1920s and early 1930s with it. I remember a guy named Richardson. He thought he was better than we were because his ancestors had not been in slavery. He was born in Central America. His ancestors had not been slaves, and mine had. We had a terrible discussion at Morris Brown with him. Some of us got sort of angry over it.

R: *November 8, 1932, Roosevelt elected president. (413)*

JRR: During Hoover's presidency, we had gone down to where everybody was tired of government. President Roosevelt was elected, and he started the NRA, TWA, and WPA, the CCC, and all of those forms of activities to give folk jobs and help out folk's situations. He started social security and all like that.

R: *Hitler. (417)*

JRR: We read about what he did. And of course, he was chancellor. People in some circles said that he thought he was god. When he was elected chancellor, he was supposed to be serving people, but he got the group to serve him to the extent that they had to address him Hiel Hitler.

R: *April 1933, FDR took the dollar off the gold standard. (419)*

JRR: It put more money into circulation. Everybody was getting a job and getting money. That was when men took their shovels and went out and cleaned out the creeks and things, with cigars in their mouths.

There was a CCC camp at Waco. That is where Margaret met her husband, Ralph. There was another fellow in the camp who liked Margaret, and her mother preferred him to Ralph because Ralph was not as tall as Margaret.

R: *Jesse Owens, 1936 Olympics.* (458)

JRR: Hitler failed to recognize him because he outran the German contenders. There was a picture of him in the paper, winning the Olympics.

R: *Joe Louis, 1937.* (471)

JRR: People would go into a café and say, "I want a steak, *beat* up like Joe Louis beat up Max Baer." I lived in what you may call rural Georgia, and it was a long time before things reached rural Georgia.

R: *Radio.*

JRR: Momma had abandoned a radio, and they gave it to me. I bought a battery for it, and it wouldn't work. Finally, I bought a radio. It was the first one that I bought. We were living on Connecticut Avenue in Tallapoosa. That was a time Eugene Talmadge was governor of Georgia. I turned my radio on one morning—they announced that Eugene Talmadge was dead.

R: *Minimum wage 40¢ an hour, June 25, 1938.* (483)

JRR: It had no impact on teachers' salaries. About 1934, men were working for fifty cents a day. Some Black women walked two and a half miles, six days a week to work for fifty cents a week in domestic service. I remember when the minimum wage went from a dollar to $2.25 and up.

R: *James Weldon Johnson, June 1938.* (483)

JRR: He was a poet.

R: *Kate Smith.* (487)

JRR: She sang "God Bless America."

R: *Marian Anderson denied DAR Hall admission, February 1939.* (490)

JRR: She sang "Ave Maria."

R: *U.S. Interns 100,000 Japanese-Americans in April 1942.* (535)

JRR: They had granted them citizenship. Then after Pearl Harbor, they moved them back about fifty miles from the shore in California and put them in concentration camps. I don't know what segment of the community was ashamed by that. Because if you grant me citizenship, then I am supposed to have it.

R: *Reparations for blacks.*

JRR: I don't see how they can do much about that because they have to have certain records. You see, I had paid ten cents a week on you ever since you were born until last year to the Life Insurance Company of Georgia [52 x 68 x $.10 = $353.60]. And on all the rest of our children up until this year [2001]. They accepted my money but cut me off and sent me about sixty-one dollars, the highest amount in settlement for each one. I had paid a quarter over the same time for Mama, and I didn't get but $488 for her life insurance [52 x 68 x $.25 = $884.00].

R: *Selective Service, November 1942.*

JRR: I registered. I had a registration card in my wallet, the last time I saw my wallet. I was classified 4F. Then I was classified 1A. I could be drafted at any time. I remember Mama cried because I may have to leave her with all of those children. All of you all were born by then.

R: *George Washington Carver.*

JRR: R. R. Moten was from Tuskegee. When he came to Forsyth to speak, his topic was "George Washington Carver" and what he had done with the peanut, sweet potato, and those things.

R: *Franklin D. Roosevelt, February 1945. (584)*

JRR: It was not generally known that he had a mistress in Warm Springs, Meriwether County, although we knew he had a home there for his polio health.

R: *April 12, 1945, Roosevelt Died. (588)*

JRR: Reverend Willis and I were discussing when Truman became president. We were accepting President Truman because he had been chosen by Franklin Roosevelt. Reverend Willis had that much confidence in President Roosevelt. Truman created the expression "The buck stops here!" It was an outstanding expression of him and his presidency. He was a good president.

R: Did the job make the man?

JRR: I would be leaning that way.

R: It has a heck of a challenge with the current president.

JRR: Oh boy! I was wondering the other day how history will treat Bush. Of course, it is yet to come, what Bush will do to himself and to the country.

R: *Nuremberg War Crimes, November 1945. (602)*

JRR: I remember that they were trying some of those who had become "big shots" in the army and to decision making in the country. I don't recall any names.

R: *In November 1948, Truman won over Dewey. (653)*

JRR: We were having our race differences.

R: *Frank Sinatra, popular in 1949. (667)*

JRR: I liked his recordings of songs.

R: *Jackie Robinson, 1949. (667)*

JRR: I was aware of him and expected that he might be envied by many whom he would show up by his superior playing skills. He would give it all he had.

R: *George Wallace, April 1963. (897)*

JRR: I know who George Wallace was and what his philosophy of life was, but I never did see him in person. He was shot in your neighborhood in Maryland. He admitted some things that I didn't think he would ever admit.

R: *Malcolm X, May 1963. (898)*

JRR: I didn't know anything about him. There was so much going on at the height of the race question.

R: *Martin Luther King, Jr., August 1963. (902)*

JRR: I knew about him and had to lecture about him. I was invited to a White church out here about ten miles from Cornelia. It was the Sunday before he was killed (March 31, 1968). We had a panel discussion. There were Whites on the panel, and I was there defending Martin Luther King.

I recall when they asked that question. We had been sitting down answering questions. I told them that I had better stand up to answer this. Martin Luther King at that time was a dangerous item to discuss. He had given the "I have a dream" speech in August 28, 1963, that had reached a lot of folk.

A White lady asked the question. I got up to answer her question, which I can't recall specifically. I said that we don't know the situation there, but we know the situation here. And wouldn't it be nice if all of God's children could walk together hand in hand and down the streets of Cornelia and treat each other like brothers? That is what the Constitution promised. I quoted the Constitution like that. They applauded when I sat down. I remember it was a sort of explosive subject at the time.

R: *John F. Kennedy killed shortly after noon, Friday, November 22, 1963. (906)*
JRR: I don't recall exactly what I was doing. If it was a weekday, I was teaching. I was looking at television when Lee Harvey Oswald was killed by Jack Ruby [Sunday, November 24, 1963].
R: *Barry Goldwater Nominated for President by GOP, July 15, 1964. (918)*
JRR: I didn't like him. That was when I fell out with the Republican Party. There was a race situation, and Barry Goldwater said, "Deep down in your heart you know you have a better [something] than a Black man." Before that time, I had voted with the Republicans, most of the time.

I have not taken the oath of office to be either Democrat or Republican. Deep down, however, I am a dyed-in-the-wool Democrat. When I looked at the Republicans, I went along with them generally because of Lincoln's Emancipation Proclamation. He signed it into law, and he was a Republican. That is all I had to go on. Nothing else about them satisfied me. I didn't see anything that led me to believe that he was in favor of the Blacks or freeing slaves.

R: *Cassius Clay, Muhammad Ali, Tuesday, May 25, 1965, knocked out Sonny Liston in 48 seconds. (934)*
JRR: He was a boxer, and he went to the Olympics, representing the United States. He held up his fist when he was not supposed to or something like that.
R: *Adlai Stevenson died July 14, 1965.*
JRR: He was running for president against Eisenhower but did not get elected. He was a might too liberal.
R: *Ronald Reagan.*
JRR: Someone asked him what he was going to do about the race problem. He said, "I didn't know there was one."
R: *Vietnam War.*
JRR: It appeared to me that they were trying to do the same thing the French tried to do and failed. I was on the draft board for Habersham County during the

Vietnam War. My appointment came from Washington DC and was signed by President Nixon. It was a federal function. Rather than be drafted, some men left the country and went to Canada.

R: *Nixon was President in June 1969, Elected in November 1968 (with Agnew as Vice President). (992)*

JRR: He was elected for a second term in November 1972. He ran against McGovern. Watergate was Nixon's downfall.

R: *Gerald Ford became Vice President in October 1973. August 8, 1974, Nixon quit. (1082)*

JRR: Ford appointed Rockefeller as vice president.

Chapter 31

THE HUMAN CONDITION

May 27, 2001
This conversation dealt with some of Dad's social perspectives, what he had observed about the human condition, and what prophecy it might have for the twenty-first century. The topics were previously agreed upon by both father and son.

Robert: *The presidential election of November 2000.*
JRR: I feel like the Republicans stole the election outright. Until the election, I thought the Supreme Court stood for law and order, equal rights for all and injustice to none. This showed me that they were party oriented as well because they were appointed by a Republican administration.

    I was really surprised at the justices being partisan. They want justice for the party that nominated them to the Supreme Court. I thought the Supreme Court ought to have justice for all, equal rights to everybody, and special privilege to nobody; but they didn't show that in the 2000 presidential election issue.

    I lost respect for the Supreme Court. I don't know what Bush will do, but his doings will be for the "good of the party" before "the good of the country." He reminds me of Barry Goldwater.

R: *The electorate.*
JRR: The people who voted for Bush across party lines in this election, were voting more *against* Clinton, than *for* Bush. Too, many have memories of Papa Bush's years in office. Most of them rely too heavily on "party." Gore could not distance himself far enough away from Clinton.

R: *Justice Clarence Thomas.*
JRR: He asked no questions during the court's consideration of the election issue. Asked, "If you don't get to the question stage, where are you?" "Just rambling around."

    I think he made passes at that woman, and she resented it. Of course, I'm not a judge or jury, but I think he made a pass at her.

R: *Education records.*

JRR: From the State Department of Education on down, they don't keep good records. I'm sure they have shredded their policy on teacher salaries in Georgia. At first, they had one for the amount of education, two years of college. They upgraded to three years. They upgraded Blacks with similar training who got 60 percent of that salary. See the book on Habersham County.[1]

R: *The evolution in discipline of children and youths.*

JRR: It has been almost a revolution since I was a boy and since my sons were boys. I came along when they could chastise the children, and now it's not chastising—it's just abuse. I used to get whippings. My parents used to whip things out of me.

Discipline needs redefining. I think with a child you have to deal with him in a language that he understands. For many, spanking the child until he got the understanding that he shouldn't do this, that, or the other was effective. Now, you better not spank a child or you may get in trouble with the law.

R: *Professional sports.*

JRR: I know something has to give soon. In the field of athletics, the pay has just gone sky-high and just exploded, and something has got to be done. It takes a rich man to go to see a football game. I think the whole country is out of shape. It got that way by the agents. They pooled their players, and one accepted him as an agent, and they sell them to the highest bidder. That is what messed up the game.

No player is worth that much money. The buck has to stop somewhere. I don't know how it can go on like it is. They ought to be able to make a decent living, but most of these salaries are above most players' capacity to handle. Those fellows like Rocker have a "million-dollar arm and a ten-cent head." They get into trouble. Money talks. Everybody will cater to you.

R: *Church membership of men in decline.*

JRR: It might have been a mistake on my part, not doing anything but teaching Sunday school. As we are now, we've got a situation. I'm in a little church. They don't have any other men, except two[2]: one has the mind of an imbecile and the other a moron. And that is who they have to send as a delegate and to certain places as a representative.

If I had accepted a stewardship or deaconship or anything, they would have made me that, and I guess I would have been different. But since I didn't, I am glad I didn't. For one thing, I don't have to worry about going to hell for misplacing church money and all like that, like so many folk.

R: *Churches and money.*

JRR: You know about the Southern Baptist Convention. As with athletics, churches are affected. I just think you need to wipe the slate clean and start over. The preacher who was the head of the Southern Baptist used the church's funds and bought a house for a woman who was not his wife. He had a "mare's nest." That means a heap of money there for the taking. I thought he got off without making time, but I understand he's serving time in federal prison.

R: *Spy Plane in China.*

JRR: It seems that China wants to get rid of it. We want to get it. They don't want the US to fix it up to fly it out. They want them to take it apart and ship it out. They want to examine all the parts before they take them out of China. I can't read much now. I read the headlines and see and hear the news on television, mostly.

R: *Aging and disabilities.*

JRR: I can't look forward to anything now, I hurt so badly, that I can't think about somebody else. I can hardly stand up. I can't get around too well. In fact, I hate to try to go to church—my legs hurt so badly. We don't have any way at my church for the wheelchair to go in. My eyesight is going. That makes reading difficult. Hearing is dim. You hate to ask everybody to speak up.

If you can't hear, you don't understand if your point was made. You feel as if you are up on a shelf. As you know, driving all these years, I did not hurt anyone. When my vision got poor, I quit driving. I don't enjoy the service like I used to. It doesn't mean a thing if you can't hear. It's like "God so loved the world" and "I like apple pie." There is no connection.

Endnotes

1. Sisk, Betty & Gowder, Ellene (1995). *Once Upon a Time: Schools of Habersham County Georgia.* J & M printing, Tiger, GA 30576
2. Archaic terms for the mentally challenged

Chapter 32

LESSONS LEARNED

1. Make sure the world has one less rascal in it by your actions.

Audrey: Dad, I have heard you say this since I was a little girl. I understand that it has to do with being a good person rather than one who behaves in a despicable manner.

JRR: That's right. You're supposed to be responsible for your actions and the *consequences* of those actions.

A: This journal[1] has some suggestions for advice you can leave for your descendants. This section has to do with *lessons*. Would you like to hear these and maybe answer them, if that is what you'd like to say?

JRR: Yes, that would be one way of getting at it.

A: "Recall five of the most important lessons you learned in life." (25)

JRR:
1. Mind your own business.
2. Respect the rights of others. I didn't try to run over anybody. Everybody has rights.
3. "Be prepared." I learned to be prepared. I learned it late in life.
4. Tell the truth.
5. Don't make a statement without having a reason for it. That is why I adopted plain geometry as my pet subject. I heard a man say it was the finest training a mind could get because it didn't let you make a statement without a reason.
6. The golden rule. I try not to trespass against anybody.

A: "Share your idea of what makes a good friend" (41)

JRR: Well, I guess the golden rule: "Do unto others as you would have them do unto you."[2] And I guess honesty and truth because I hate for anybody to tell me a lie. I try not to trespass against anybody.

A: "Share some of your insights from working with others" (57).

JRR: *One. I advocated telling the truth.* If you can't tell the truth on anybody, don't say anything. Don't say anything just to get rid of somebody. Let your "yea" be yea and your "nay" be nay. Put things on the table and thrash them out.

Two. I always respected the rights of my teachers and the rights of my students. I respected them like I did my own children. I had some teachers I didn't like, but for the sake of peace, I had to tolerate them. Like Clemmie Adams didn't know how to teach, but her mother was a pillar of the community. And I couldn't do anything about it.

A: "Share your ideas of what it takes for a husband and wife to maintain a healthy marriage" (105).

JRR: The husband: First, keep the Boy Scout oath: "On my honor I will do my best to do my duty to God and my country and . . . to help other people at all times and to keep myself physically strong, mentally awake, and morally straight." Second, keep the marriage vow: "Forsake all others and cleave to your wife." And let her be a princess. The wife: Keep the marriage vow. Forsake all others and be true to your husband.

2. In your dealings with others, treat them fairly.

A: Would you give me an example that you experienced but had not mentioned before.

JRR: There are a lot of them and most of them have to do with money. You may wonder how Bob Rosser got the land he owned. He had rented that land because he was born on it. The Cochrans had taken it for taxes. They moved away and had it up for sale. He bought it, along with Mamie's help.

The old man got scared about his deed and gave it to me to keep. When he got ready to get his pension, he didn't have any better sense than to tell them he sold it to me. They cut him off, and he sent Joe up here to get it, and he carried it to Greenville to straighten that out.

He wanted to have enough money to leave each of his children two hundred dollars apiece. He sold trees for pulpwood. Joe, Ellis, and JI cut down all the pulpwood on the property. It was 202 acres of land. I never wanted to know how much he got because I never accepted any money from him except the ten cents for train fare when I had to go back to school.

When Pa died, Mamie and Joe demanded fifty acres apiece. That would have left one hundred two acres to go to the rest of us. I told them to leave me out. Jennie Mae told them to leave her out. I think Laura didn't want to be a part of it. JI did not like it, and things remained foreign between him and the two of them. I don't know how the rest of them felt: Elmira, Amy, Ellis, William, and Rofford; but they couldn't settle it peacefully. Mamie ended up with all except fifty acres, which went to Joe, and then to his widow, Bessie.

Now, all of Pa's children except Rofford and me are gone. According to Dorothy, Jennie Mae's daughter, Mamie signed over the 152 acres to her husband, Alfrez. He was half-crazy, and his greedy-gut relatives took it. The Shuffer relatives took that land and sold it to Marquis Grissom.

3. Keep working until you finish your education.

A: That speaks for itself.
JRR: Yes, it's tempting to stop, but if you keep going to the end, one day you'll be glad you did.

This narrative was discovered in Dad's effects. It is his Principal's Message from the last newsletter printed by the Regional High School before it was closed (The Regional Orbit, Volume 4, Number 16, May 1966).

My Dear Young Friends:

In this, my last admonition in this capacity, there are so many subjects on which I would like to give some strength. And, they are all of the "*do* or *die*" variety, so much so, until we are undecided which would arouse your curiosity most.

Since the present circumstances, situations and conditions seem to have most of you in an encircled gloom of skeptical anticipation, I must haste to tell you that you must live off your record the rest of your lives. No matter what may cause you to error, the record is made. Like death, once it occurs, it's always final.

You might start by setting yourselves codes of standards and sets of ideals, then live up to them twenty-four hours of every day, the rest of your lives. Let your codes and sets take into consideration your rights, when they end, and when the other fellow's begin. Try hard not to trespass. You should have more respect for your neighbor's property than your pet pup has.

You shouldn't let anybody make you make an idiot of yourselves. Don't waste your time looking for a fair treatment and justice. Take advantage of every opportunity that comes your way.

More than ever before in the history of mankind only 10% of one's life is, "What he makes it," and 90% of his life is, "How he takes it." Like driving on today's highways, protecting one's self is not enough. If you do not also look out for the other fellow, you will also be destroyed with him.

You will meet people who will try to have you believe that they know all and what is best to do at all times. Don't be fooled. No such person exists. You must think some for yourself. The Great Teacher left word for you to watch and pray, lest you enter into temptation. Take your time. You might live to see that time when "justice will roll down as water and righteousness as a mighty stream." Then those with "clean hands and pure hearts" will be able to live with themselves.

In conclusion, I hope all your dreams come true. But don't waste your time looking for something for nothing. Prepare yourselves for earning a living, and strive to succeed in making a life. God be with you. Amen.

Endnotes

1. Questions are from *A Father's Legacy: Your Life Story in your Own Words*, Terri Gibbs, project editor, J. Countryman, a division of Thomas Nelson, Inc., Nashville, TN 37214, (2000).
2. Matthew 7:12.

The Last Word

It was a few years after the turn of the century, in Meriwether County, Georgia, that a little Negro boy was born to a poor, sharecropping family. He was the fourth of eleven children, born to that family. They did get a doctor for that birth, because the landowner was wealthy. In that period Negro mothers could have a doctor, if her White landlord was rich. Thus, the mortality rate among Negro mothers was very high. This little boy survived. His memory goes back to age four. That little Negro boy was me.[1]

To really tell the truth, I was known as a little "nigger" boy, and that's what I thought I was. I humbled myself to it. But I wasn't like my daddy. He told me "not to ever 'spute a White man's word." I told him, "I'd 'spute his word and do something else to him too if he bothers me." I did dispute a White man's word many times. I didn't copy any of that attitude toward life from my daddy because my daddy was slavery oriented. I wasn't a humble ex-slave like he was. I felt like I was a man like the other men.

I didn't know that I could be anything else, but I wanted to be the best plow hand in the community. And I became the best one. There was a contest to see who could plow the straightest row. Clyde Lambert judged, and I won!

I was not taught I could be anything. I didn't know I could be president. I wanted to be a doctor, but one of my teachers discouraged me. I wanted to be a speaker. I thought that was great. Nobody in high school or college encouraged me to be anything.

I realized I was behind and thought I'd never catch up. I don't worry over spilled milk. My life has been a revolution from cutting wood to an electric stove. From "nigger boy" to respected man in the community, Dr. Rosser.

Nineteen seventy-six was the year Cornelia buried a time capsule. It was nearly as big as a casket, was round, and had a lid that fastened down like a casket. It is buried somewhere around the big red apple. You can go to city hall to find out exactly where it is.

The capsule was to be opened one hundred years from then. It was projected to be dug up by the city of Cornelia. Audrey's great-grand, born a few days ago, that little baby will be too old to hear his own self fart when they dig it up. Hahaha.

Anybody could put something in it. I don't know if they kept a record because some people threw a dime or some coin in it. Dr. Cuthbertson put a check in for one million dollars. I sent a letter to go in it. I wrote one to Mama and she wrote one to me. Mama's and mine are in the same envelope. I don't remember its contents, but I think it was about all my work with the Ground Observers Corp and my work with the GMAPDC.

She gave her children some advice in that letter: "Right wrongs no man, kindness wins friends, and the longest road has an end." The author is unknown, but it was in a fourth-grade reader. We had to learn that. We had to memorize every quote and poem in the book. A lot of times it didn't have who wrote it.

I remember I told her when they dig it up, we both will be passed on. I didn't know which one would pass first. I had misgivings because I didn't want to leave her and I didn't want her to leave me here. I just can't get used to Mama being gone. I want us to be together in heaven.

I'm so glad we got to take some trips together. After Mama retired we traveled a lot with the Young At Heart[2] for several years. I remember when we went to the Bahamas. The people there were Black and most of the people in our group were White and Mama looked like them. The people of the Bahamas tried to sell her their goods and didn't bother with me. Hahaha.

I think Hawaii was the best. Two who were to go were in an accident and we took their places. I had never been on a plane. I had intended to get scared but the ride was so "smooth" I forgot to be scared and enjoyed it! We flew from island to island. By the time we got back I had a lot of flights under my belt. That was about the last long trip we took together.

We knew we wouldn't be here always. We hoped that all of our children would be successful and not dependent on what we would leave them. Getting them all educated was our aim, and we gave four out of five a college education. Anything after that they had to pay for themselves. As I think about it, I may have overemphasized education to them. Three of the five went on to earn doctorate degrees.

Habersham County Medical Center
Demorest, Georgia
May 6, 2004

I thought I would live to be a hundred, but I won't. And that's all right. I think I have done all the Lord wanted me to do. In the game of life you have to "know when to hold them, know when to fold them, know when to walk away, and know when to run."[3] I've had to hold 'em. I've had to fold 'em. I've had to walk away, and I've had to run. I did the best I could with what I had. I believe I've done what the Lord wanted me to do. I hope so. I hope he's pleased with me.

He gave his son a charge. Jesus did what he had to do. When he was finished, he knew there was no need to hang around. I don't believe he expects us to do more than he expected his son to do. I think God gives all of us a charge. We have to keep at it through all the ups and downs. When we've done it, we ought to have sense enough to know when to lay it down.

"I have fought a good fight, I have finished my course, I have kept the faith."[4]

Endnotes

1. Written by JR Rosser, Sr., as a beginning of his obituary, at least twenty years before his death. Discovered in his personal papers after his death.
2. "Dr. and Mrs. Rosser were both such wonderful people, it was a pleasure to have them travel with us. They started going with us when the club began in 1977. And I know they went until the mid 1980s. I remember them going to Nashville, New Orleans, Chattanooga, Myrtle Beach, Savannah, the New England tour, the western states, the Epcot Center, the cruise to the Bahamas, Hawaii, and lots of other places close by. If we weren't going to church on Sunday morning Dr. Rosser would do a devotion for us. He had a lovely voice. He would sing and we all would sing with him. I would feel bad when they were served last, but they would say it was fine. None of the things we ran into on the trips ever bothered them. They were just delightful." (Personal communication, Mozelle Wilbanks, Travel Coordinator, the Young At Heart Club, later named the Golden Apple Club, April 2009).
3. Lyrics from "The Gambler," written and sung by Kenny Rogers.
4. Second Timothy 4:11 King James Version

Part Three

Quotations

Scripture

Most of the quotes he gave from the Holy Bible came from the Old Testament and from the King James Version. These are the four divisions of the Psalms that he quoted most frequently and would recite in entirety. They seemed to be a raison d'être, as well as to provide calm reflection of his faith in God. He relied on these and the other biblical passages for reassurance.

PSALM 1

1. Blessed *is* the man that walketh not in the counsel of the ungodly, nor standeth in the way of sinners, nor sitteth in the seat of the scornful.
2. But his delight *is* in the law of the LORD; and in his law doth he meditate day and night.
3. And he shall be like a tree planted by the rivers of water, that bringeth forth his fruit in his season; his leaf also shall not wither; and whatsoever he doeth shall prosper.
4. The ungodly *are* not so: but *are* like the chaff which the wind driveth away.
5. Therefore the ungodly shall not stand in the judgment, nor sinners in the congregation of the righteous.
6. For the LORD knoweth the way of the righteous: but the way of the ungodly shall perish.

PSALM 8

1. O LORD our Lord, how excellent *is* thy name in all the earth! Who has set thy glory above the heavens.
2. Out of the mouths of babes and sucklings hast thou ordained strength because of thine enemies, that thou mightest still the enemy and the avenger.
3. When I consider thy heavens, the work of thy fingers, the moon and the stars, which thou hast ordained;
4. What is man, that thou art mindful of him? And the son of man, that thou visitest him?
5. For thou hast made him a little lower than the angels, and hast crowned him with glory and honour.
6. Thou madest him to have dominion over the works of thy hands; thou hast put all *things* under his feet:
7. All sheep and oxen, yea, and the beasts of the field;
8. The fowl of the air, and the fish of the sea, and *whatsoever* passeth through the paths of the seas.
9. O LORD our Lord, how excellent *is* thy name in all the earth!

PSALM 23

1. The LORD *is* my shepherd; I shall not want.
2. He maketh me to lie down in green pastures: he leadeth me beside the still waters.
3. He restoreth my soul: he leadeth me in the path of righteousness for his name's sake.
4. Yea, though I walk through the valley of the shadow of death, I will fear no evil: for thou *art* with me; thy rod and thy staff they comfort me.
5. Thou preparest a table before me in the presence of mine enemies: thou anointest my head with oil; my cup runneth over.
6. Surely goodness and mercy shall follow me all the days of my life: and I will dwell in the house of the LORD for ever.

PSALM 121

1. I will lift up mine eyes unto the hills, from whence cometh my help.
2. My help *cometh* from the LORD, which made heaven and earth.
3. He will not suffer thy foot to be moved: he that keepeth thee will not slumber.
4. Behold, he that keepeth Israel shall neither slumber nor sleep.
5. The LORD *is* thy keeper: the LORD is thy shade upon thy right hand.
6. The sun shall not smite thee by day, nor the moon by night.
7. The LORD shall preserve thee from all evil: he shall preserve thy soul.
8. The LORD shall preserve thy going out and thy coming in from this time forth, and even for evermore.

THE SECOND BOOK OF MOSES, CALLED EXODUS: CHAPTER 20

3   Thou shalt have no other gods before me.
4   Thou shalt not make unto thee any graven image, or any likeness *of any thing* that *is* in heaven above, or that *is* in the earth beneath, or that *is* in the water under the earth:
5   Thou shalt not bow down thyself to them, nor serve them: for I the LORD thy God *am* a jealous God, visiting the iniquity of the fathers upon children unto the third and fourth *generation* of them that hate me;
6   And shewing mercy unto thousands of them that love me, and keep my commandments.
7   Thou shalt not take the name of the LORD thy God in vain; for the LORD will not hold him guiltless that taketh his name in vain.
8   Remember the sabbath day to keep it holy.
9   Six days shalt thou labour, and do all thy work:
10 But the seventh day *is* the sabbath of the LORD thy God: *in it* thou shalt not do any work, thou, nor thy son, nor thy daughter, thy manservant, nor thy maidservant, nor thy cattle, nor thy stranger that *is* within thy gates:
11 For *in* six days the LORD made heaven and earth, the sea, and all that in them *is*, and rested the seventh day: wherefore the LORD blessed the sabbath day, and hallowed it.
12 Honor thy father and thy mother: that thy days may be long upon the land which the LORD thy God giveth thee.
13 Thou shalt not kill.
14 Thou shalt not commit adultery.
15 Thou shalt not steal.
16 Thou shalt not bear false witness against thy neighbor.
17 Thou shalt not covet thy neighbour's house, thou shalt not covet thy neighbor's wife, nor his manservant, nor his maidservant, nor his ox, nor his ass, nor anything that *is* thy neighbor's.

THE GOSPEL ACCORDING TO ST. MATTHEW: CHAPTER 5

1. And seeing the multitudes, he went up into a mountain: and when he was set, his disciples came unto him:
2. And he opened his mouth, and taught them, saying,
3. Blessed *are* the poor in spirit: for theirs is the kingdom of heaven.
4. Blessed *are* they that mourn: for they shall be comforted.
5. Blessed *are* the meek: for they shall inherit the earth.
6. Blessed *are* they which do hunger and thirst after righteousness: for they shall be filled.
7. Blessed *are* the merciful: for they shall obtain mercy.
8. Blessed *are* the pure in heart: for they shall see God.
9. Blessed *are* the peacemakers: for they shall be called the children of God.
10. Blessed *are* they which are persecuted for righteousness' sake: for theirs is the kingdom of heaven.
11. Blessed *are* ye, when *men* shall revile you and persecute *you* and say all manner of evil against you falsely, for my sake.
12. Rejoice, and be exceeding glad: for great *is* your reward in heaven: for so persecuted they the prophets which were before you.

POETRY

Excerpts from these poems were often quoted, especially the last verses, in teaching rules for everyday living. They were inspiration by which he governed his own life. They offer insight into the drive that kept him grounded.

INVICTUS[1]

William Ernest Henley

Out of the night that covers me,
Black as the Pit from pole to pole,
I thank whatever gods may be
For my unconquerable soul.

In the fell clutch of circumstance
I have not winced nor cried aloud.
Under the bludgeonings of chance
My head is bloody, but unbowed.

Beyond this place of wrath and tears
Looms but the horror of the shade,
And yet the menace of the years
Finds, and shall find me, unafraid.

It matters not how strait the gate,
How charged with punishment the scroll,
I am the master of my fate;
I am the captain of my soul.

IF[2]

Rudyard Kipling

If you can keep your head when all about you
Are losing theirs and blaming it on you;
If you can trust yourself when all men doubt you,
But make allowance for their doubting too;
If you can wait and not be tired by waiting,
Or, being lied about, don't deal in lies,
Or, being hated, don't give way to hating,
And yet don't look too good, nor talk too wise;

If you can dream—and not make dreams your master;
If you can think—and not make thoughts your aim;
If you can meet with triumph and disaster
And treat those two imposters just the same;
If you can bear to hear the truth you've spoken
Twisted by knaves to make a trap for fools,
Or watch the things you gave your life to broken,
And stoop and build 'em up with worn out tools;

If you can make one heap of all your winnings
And risk it on one turn of pitch-and-toss,
And lose, and start again at your beginnings
And never breath a word about your loss;
If you can force your heart and nerve and sinew
To serve your turn long after they are gone,
And so hold on when there is nothing in you
Except the Will which says to them: "Hold on";

If you can talk with crowd and keep your virtue,
Or walk with kings—nor lose the common touch;
If neither foes nor loving friends can hurt you;
If all men count with you, but none too much;
If you can fill the unforgiving minute
With sixty seconds' worth of distance run—
Yours is the Earth and everything that's in it,
And—which is more—you'll be a Man my son!

EQUIPMENT[3]

Edgar A. Guest

Figure it out for yourself, my lad,
You've all that the greatest of men have had,
Two arms, two hands, two legs, two eyes
And a brain to use if you would be wise.
With this equipment they all began,
So start for the top and say, "I can."

Look them over, the wise and great
They take their food from a common plate,
And similar knives and forks they use,
With similar laces they tie their shoes.
The world considers them brave and smart,
But you've all they had when they made their start.

You can triumph and come to skill,
You can be great if you only will.
You're well equipped for what fight you choose,
You have legs and arms and a brain to use,
And the man who has risen great deeds to do
Began his life with no more than you.

You are the handicap you must face,
You are the one who must choose your place,
You must say where you want to go,
How much you will study the truth to know.
God has equipped you for life, but He
Lets you decide what you want to be.

Courage must come from the soul within,
The man must furnish the will to win.
So figure it out for yourself, my lad.
You were born with all that the great have had,
With your equipment they all began,
Get hold of yourself and say: "I can."

A PSALM OF LIFE[4]

Henry Wardsworth Longfellow

Tell me not in mournful numbers,
Life is but an empty dream!
For the soul is dead that slumbers,
And things are not what they seem.

Life is real! Life is earnest!
And the grave is not its goal;
Dust thou art, to dust returnest,
Was not spoken of the soul.

Not enjoyment, and not sorrow,
Is our destined end or way;
But to act, that each tomorrow
Find us farther than today.

Art is long, and time is fleeting,
And our hearts, though stout and brave,
Still, like muffled drums, are beating
Funeral marches to the grave.

"In the world's broad field of battle,
In the bivouac of Life,
Be not like dumb, driven cattle!
Be a hero in the strife!"

Trust no Future, howe'er pleasant!
Let the dead Past bury its dead!
Act, act in the living Present!
Heart within, and God o'erhead!

Sharecropper to Landowner

Lives of great men all remind us
We can make our lives sublime,
And, departing, leave behind us
Footprints on the sands of time;

Footprints, that perhaps another,
Sailing o'er life's solemn main,
A forlorn and shipwrecked brother,
Seeing, shall take heart again.

Let us, then, be up and doing,
With a heart for any fate;
Still achieving, still pursuing,
Learn to labor and to wait.

THE VILLAGE BLACKSMITH[5]

Henry Wadsworth Longfellow

Under a spreading chestnut tree
The village smithy stands;
The smith, a mighty man is he,
With large and sinewy hands;
And the muscles of his brawny arms
Are strong as iron bands.

His hair is crisp, and black, and long,
His face is like the tan.
His brow is wet with honest sweat,
He earns whate'er he can,
And looks the whole world in the face,
For he owes not any man.

Week in, week out, from morn till night,
You can hear his bellows blow;
You can hear him swing his heavy sledge
With measured beat and slow,
Like a sexton ringing the village bell,
When the evening sun is low.

And children coming home from school
Look in at the open door;
They love to see the flaming forge,
And hear the bellows roar,
And watch the burning sparks that fly
Like chaff from a threshing-floor.

He goes on Sunday to the church,
And sits among his boys;
He hears the parson pray and preach,
He hears his daughter's voice,
Singing in the village choir, and it makes his heart rejoice.

Sharecropper to Landowner

It sounds to him like her mother's voice,
Singing in Paradise!
He needs must think of her once more,
How in the grave she lies;
And with his hard, rough hand he wipes
A tear out of his eyes.

Toiling,—rejoicing,—sorrowing,
Onward through life he goes;
Each morning sees some task begin
Each evening sees it close;
Something attempted, something done,
He earned a night's repose.

Thanks, thanks to thee, my worthy friend,
For the lesson thou hast taught!
Thus at the flaming forge of life
Our fortunes must be wrought;
Thus on its sounding anvil shaped
Each burning deed and thought!

CASEY AT THE BAT[6]

Ernest Lawrence Thayer

The outlook wasn't brilliant for the Mudville nine that day;
The score stood four to two, with but one inning more to play,
And then when Cooney died at first, and Barrows did the same,
A pall-like silence fell upon the patrons of the game.

A straggling few got up to go in deep despair. The rest
Clung to that hope which springs eternal in the human breast;
They thought, "If only Casey could get a whack at that—
We'd put up even money now, with Casey at the bat."

But Flynn preceded Casey, as did also Jimmy Blake,
And the former was a hoodoo, while the latter was a cake;
So upon that stricken multitude grim melancholy sat;
For there seemed but little chance of Casey getting to the bat.

But Flynn let drive a single, to the wonderment of all,
And Blake, the much despised, tore the cover off the ball;
And when the dust had lifted, and men saw what had occurred,
There was Jimmy safe at second and Flynn a-hugging third.

Then from five thousand throats and more there rose a lusty yell;
It rumbled through the valley, it rattled in the dell;
It pounded on the mountain and recoiled upon the flat,
For Casey, mighty Casey, was advancing to the bat.

There was ease in Casey's manner as he stepped into his place;
There was pride in Casey's bearing and a smile on Casey's face.
And when, responding to the cheers, he lightly doffed his hat,
No stranger in the crowd could doubt 'twas Casey at the bat.

Ten thousand eyes were on him as he rubbed his hands with dirt.
Five thousand tongues applauded when he wiped them on his shirt.
Then while the writhing pitcher ground the ball into his hip,
Defiance flashed in Casey's eyes, a sneer curled Casey's lip.

And now the leather-covered sphere came hurtling through the air,
And Casey stood a-watching it in haughty grandeur there.
Close by the sturdy batsman the ball unheeded sped—
"That ain't my style," said Casey. "Strike one!" the umpire said.

From the benches, black with people, there went a muffled roar,
Like the beating of the storm-waves on a stern and distant shore;
"Kill him! Kill the umpire!" shouted someone on the stand;
And it's likely they'd have killed him had not Casey raised his hand.

With a smile of Christian charity great Casey's visage shone;
He stilled the rising tumult; he bade the game go on;
He signaled to the pitcher, and once more the dun sphere flew;
But Casey still ignored it, and the umpire said, "Strike two!"

"Fraud!" cried the maddened thousands, and echo answered "Fraud!"
but one scornful look from Casey and the audience was awed.
They saw his face grow stern and cold, they saw his muscles strain,
And they knew that Casey wouldn't let that ball go by again.

The sneer has fled from Casey's lip, the teeth are clenched in hate;
He pounds with cruel violence his bat upon the plate.
And now the pitcher holds the ball, and now he lets it go,
And now the air is shattered with the force of Casey's blow.

Oh, somewhere in this favored land the sun is shining bright,
The band is playing somewhere, and somewhere hearts are light,
And somewhere men are laughing, and little children shout;
But there is no joy in Mudville—mighty Casey has struck out.

SEPTEMBER[7]

Helen Hunt Jackson

The golden-rod is yellow;
The corn is turning brown;
The trees in apple orchards
With fruit are bending down.

The gentian's bluest fringes
Are curling in the sun;
In dusty pods the milkweed
Its hidden silk has spun.

The sedges flaunt their harvest,
In every meadow nook;
And asters by the brook-side
Make asters in the brook.

From dewy lanes at morning
The grapes' sweet odors rise;
At noon the roads all flutter
With yellow butterflies.

By all these lovely tokens
September days are here,
With summer's best of weather,
And autumn's best of cheer.

But none of all this beauty
Which floods the earth and air
Is unto me the secret
Which makes September fair.

'Tis a thing which I remember;
To name it thrills me yet;
One day of one September
I never can forget.

THE BRIDGE BUILDER[8]

Will Allen Dromgoole

An old man, going a lone highway,
Came, at the evening, cold and gray,

To a chasm, vast, and deep, and wide,
Through which was flowing a sullen tide.

The old man crossed in the twilight dim;
The sullen stream had no fears for him;

But he turned, when safe on the other side,
And built a bridge to span the tide.

"Old man," said a fellow pilgrim near,
"You are wasting your strength with building here;

Your journey will end with the ending day;
You never again will pass this way;

You have crossed the chasm, deep and wide—
Why build you a bridge at the eventide?"

The builder lifted his old gray head:
"Good friend, in the path I have come," he said,

"There followeth after me today
A youth, whose feet must pass this way.

This chasm, that has been naught to me,
To that fair-haired youth may a pitfall be.

He, too, must cross in the twight dim;
Good friend, I am building the bridge for him."

HOUSE BY THE SIDE OF THE ROAD[9]

Samual Walter Foss

There are hermit souls that live withdrawn
In the place of their self-content;
There are souls like stars, that dwell apart,
In a fellowless firmament;
There are pioneer souls that blaze the paths
Where highways never ran—
But let me live by the side of the road
And be a friend to man.

Let me live in a house by the side of the road
Where the race of men go by—
The men who are good and the men who are bad,
As good and as bad as I.
I would not sit in the scorner's seat
Nor hurl the cynic's ban—
Let me live in a house by the side of the road
And be a friend to man.

I see from my house by the side of the road
By the side of the highway of life,
The men who press with the ardor of hope,
The men who are faint with the strife,
But I turn not away from their smiles nor their tears,
Both parts of an infinite plan—
Let me live in my house by the side of the road
And be a friend to man.

I know there are brook-gladdened meadows ahead,
And mountains of wearisome height;
That the road passes on through the long afternoon
And stretches away to the night.
And still I rejoice when the travelers rejoice
And weep with the strangers that moan,
Nor live in my house by the side of the road
Like a man who dwells alone.

Sharecropper to Landowner

Let me live in a house by the side of the road,
Where the race of men go by—
They are good, they are bad, they are weak, they are strong,
Wise, foolish—so am I.
Then why should I sit in the scorner's seat,
Or hurl the cynic's ban?
Let me live in my house by the side of the road
And be a friend to man.

STRAIGHTEN UP AND FLY RIGHT[10]

Ephraim David Tyler

Just a few brief words of comment, a suggestion, an advice
Which might prove a precious jewel at a very meager price.
I am speaking to the nation and to all men far and near.
People who can think and reason should incline a listening ear.
Sex and color will not matter, male or female black or white;
As the monkey told the buzzard,
    "Straighten Up and Fly Right."

This is just a slang expression, a "Rag Song," which screen stars sing,
But it carries a deep moral for the peasant or the king.
As a rule I use expressions which will rank among the best,
And I only quote the authors whose quotations stand the test.
But no man can fly an airplane like a school boy flies his kite—
Let the soldiers and civilians
    "Straighten Up and Fly Right."

Once again may I remind you that this world is STILL in war.
We may shortly face conditions which this nation never saw.
Doubtless, some of us have suffered. We may have to suffer more,
We will make some sacrifices which we never made before.
Don't let any person fool you. This is everybody's fight.
All of us will need some War bonds—
    "Straighten Up and Fly Right."

All the food administrators who compose our OPA,
And the men who go to Congress should know what to do and say,
We must have some sane, wise leaders to conduct our politics;
One bad move may cause the nation to be in an awful fix.
Strikes should not be tolerated. Let's keep busy day and night.
We have but one chance to conquer—
    "Straighten Up and Fly Right."

Saving souls is the great mission to be fostered by the church;
Help the sick, the blind, the helpless and the fellow on his crutch.
Try to find some wise, true leader whom the Lord has called to lead,
And be mindful of "gold-diggers," filled with selfishness and greed.
This old world is still in darkness, and the man who brings the light
Should be filled with the Christ Spirit.
    "Straighten Up and Fly Right."

You should "Straighten Up," school teachers. Ev'ry teacher in the school
Should teach ev'ry body's children to observe the Golden Rule.
"Straighten Up," you undertakers. Though we want to see you rise,
You are charging too much money for your funeral merchandise.
Straighten, straighten, all you merchants, when you go to make your flight
"Straighten Up" your moral standards.
     "Straighten Up and Fly Right."

Straighten, doctors, farmers, work shops. "Straighten Up" you food cafes.
We must straighten up this nation in at least a thousand ways.
"Straighten Up" your transportation on the land, the sea and air.
Straighten, Capital and Labor. Straighten, straighten ev'rywhere.
This old world will have to straighten or be left in a sad plight.
Straighten, straighten ev'rybody.
     "Straighten Up and Fly Right."

Endnotes

1. "Invictus" (Unconquered). http://en.wikipedia.org/wiki/Invictus
2. "If." http://www.everypoet.com/archive/poetry/Rudyard_Kipling/kipling_if.htm
3. "Equipment." http://www.tuskegee.edu/Global/story.asp?S=1626687
4. "A Psalm of Life." http://www.potw.org/archive/potw232.html
5. "The Village Blacksmith." http://www.appaltree.net/aba/poem.htm
6. "Casey at the Bat." http://ops.tamu.edu/x075bb/poems/casey.html
7. "September." http://www.poemhunter.com/poem/september-2/
8. "The Bridge Builder." http://www.essentia.com/book/poems/bridgebuilder.htm
9. "House by the . . ." http://www.ipoet.com/ARCHIVE/ORIGINAL/foss/House.html
10. Tyler, E. D. (1923). *Poems of Every Day Life,* [Publishing Company undetermined], Shreveport, Louisiana, pages 42-43. (Private Library, Leslie Jones)

Note: Tyler was an African American born in Louisiana in 1884. He was named Poet Laureate by Gov. Earl K. Long in 1951 and was affectionately called the "rustic poet." His poems about Black lifestyle were advocates for citizenship, justice, and love.

Advice

These quotes of unknown origin were spontaneously delivered, always apropos, and generally not preceded or followed by any other words. They stood alone, as advice, opinion, and resolution of the circumstance being discussed or experienced.

Four things a man must learn to do,
If he would make his calling sure:
To think without confusion, clearly;
Act from earnest motive purely;
Love his fellow man sincerely; and
Trust in God and heaven securely.

Once the labor has begun
Never leave it till it's done.
Be the labor great or small
Do it well or not at all!

"I love you mother," said little Nell.
"I love you more than tongues can tell."
Then off she went to the garden swing
Leaving her mother the wood to bring.

There's so much bad in the best of us,
And, so much good in the worst of us,
Till it hardly behooves the most of us
To talk about the rest of us.

He who knows not and knows not that he knows not is a fool, shun him.
He who knows not and knows that he knows not is simple, teach him.
He who knows and knows not that he knows is asleep, wake him.
He who knows and knows that he knows is wise, follow him.

If your lips
Would keep from slips
Five things observe with care:
To whom you speak,
Of whom you speak,
And how, and when, and where.

One-Liners

Strike with all your might while the iron is red.
Creep, terrapin, you know your gait!
Every tub has to set on its own bottom.
Sometimes you have to saw wood and say nothing.
Sooner or later, somehow or other, it all comes around in the end.
Chickens always come home to roost.
Every dog has his day.
Bad news will beat you home.
She'll rob a crippled tumblebug and put a blind man on the wrong road home.
There's never a good reason to be rude to somebody, even if you know they're wrong.

Lyrics

CAUTION: Watch out for people who act like this
"If you see me coming better step aside. A lot of men didn't and a lot of men died" ("Sixteen Tons" written by Merle Travis, recorded by Tennessee Ernie Ford).

MISSION: Do what you can to help others.
"You've got to run to the harvest field, to the harvest field, you've got to run. You've got to run to the golden yield, to the golden yield, you've got to run" (chorus of a traditional gospel melody "You've Got to Run," writer and singer unknown).

ASSURANCE: Trust in God.
"Amazing grace! How sweet the sound that saved a wretch like me! I once was lost, but now am found; was blind, but now I see . . . . Through many dangers, toils, and snares, I have already come; 'tis grace hath brought me safe thus far, and grace will lead me home" (written by John Newton).

Printed in the United States
151924LV00009B/80/P